Library of
Davidson College

IN THE SPIRIT OF ENTERPRISE
FROM THE ROLEX AWARDS

IN THE SPIRIT OF ENTERPRISE
FROM THE ROLEX AWARDS

EDITED BY
GREGORY B. STONE

Foreword by
Gerard Piel,
Publisher, *Scientific American*

Preface by
André J. Heiniger,
Managing Director, Montres Rolex S.A.

W. H. Freeman and Company
San Francisco

Photographs and art not credited below were submitted by the entrants.

Pages 16, 20, 38, 78, 96, 122, 148, 176, 192, 222, 236, 238, 255, 256, 274, 286, 300: Dale Johnson. Pages 12, 58, 128, 130, 180, 182, 260: J. Walter Thompson. Page 6: Dr. George Kennedy. Page 44: Ken Balcomb, Orca Survey, Moclips Cetological Society. Page 56: San Diego Zoo Photo. Page 86: Stern/Black Star. Page 92: Carl Frank/Photo Researchers, Inc. Pages 108, 278: Kenneth E. Lucas/Steinhart Aquarium. Page 136: John M. Burnley/Photo Researchers, Inc. Page 158: Englebert/Photo Researchers, Inc. Page 168: Nathan W. Cohen/Herpetofaunacolor. Page 200: C. J. Anderson/American Museum of Natural History. Page 208: Biblioteca Ambrosiana, Milan from Art Reference Bureau. Page 214: L.L.T.Rhodes/© Animals Animals, 1975. Page 244: Hale Observatories. Copyright © by the California Institute of Technology and The Carnegie Institution of Washington. Reproduced by permission from the Hale Observatories. Page 264: Al Grotell. Page 268: E. Hanumantha Rao/Photo Researchers, Inc. Page 278: Kenneth E. Lucas/Steinhart Aquarium. Page 292: George Holton/Photo Researchers, Inc. Page 324: Raymond A. Mendez/© Animals Animals, 1976. Page 328: Eric Simmons/Stock, Boston, Inc.

Library of Congress Cataloging in Publication Data

Main entry under title:

In the spirit of enterprise from the Rolex Awards.

"This book presents 131 of the projects submitted to 'The Rolex Awards for Enterprise' conducted by Montres Rolex S.A. in 1976."

   1.   Inventions—Awards.   2.   Science—Awards.
I.   Stone, Gregory B.   II.   Rolex Watch Company, ltd., Geneva.
T49.5.I48   001.4′3   78-14374
ISBN 0-7167-1034-X

Copyright © 1978 by Montres Rolex S.A.

No part of this book may be reproduced by any mechanical, photographic, or electronic process, or in the form of a phonographic recording, nor may it be stored in a retrieval system, transmitted, or otherwise copied for public or private use, without written permission from the publisher.

Printed in the United States of America

1 2 3 4 5 6 7 8 9

CONTENTS

Rolex Laureates are indicated by a gold crown (♛) and Honorable Mention winners by a black crown (♛).

FOREWORD, GERARD PIEL xv

PREFACE, ANDRÉ J. HEINIGER xvii

INTRODUCTION, GREGORY B. STONE xix

♛ OVER THE SOUTH POLE BY BALLOON 1
ROLAND READING PARSONS

SAVE THE ORCHIDS 7
GUALTIERO GIOVANDO

♛ THE WORLD'S LARGEST AND OLDEST ART COLLECTION 11
LUC JEAN-FRANÇOIS DEBECKER

HEARING THROUGH THE SKIN 17
DAVID FRANKLIN

♛ A SMALL, TOUGH HOVERCRAFT FOR BIG, TOUGH RIVERS 23
MICHAEL EDWIN COLE

EXPERIMENTAL FREE STATES 27
CHRISTIAN HENRIK SNELLMAN

FLYING HIGHER THAN EVER BEFORE 33
DARRYL GEORGE GREENAMYER

 THE ABOU KIR DRAIN—POLLUTION COMES TO DEVELOPING COUNTRIES 39
SAMIA GALAL SAAD

DIVING WITH KILLER WHALES 45
MARK CHRISTIAN OVERLAND

TRACKING ONE OF THE WORLD'S GREATEST
UNDERGROUND RIVERS 51
JEAN-FRANÇOIS PERNETTE

CAPTIVE BREEDING—LIFELINE TO THE FUTURE FOR
ENDANGERED SPECIES 57
BILLY LEE LASLEY

WE NEED PORTRAITS, NOT JUST PHOTOS, OF
VANISHING CULTURES 61
LUNDA HOYLE GILL

AN INTERNATIONAL COLOR STANDARD FOR
BIOLOGY 65
J. HOWARD FRANK

RETRIEVING A WORLD FROM THE STONE AGE 71
JORGE SAMUEL MOLINA BUCK

BEFORE OIL, THE TRAIL LED TO INCENSE 79
HUGO FRANZ BAUR

AN 80-FOOT STEEL CATAMARAN: ONE-MAN CRAFT
FOR ONE-MAN CREW 83
LESLIE GEORGE THOMPSON

LOOKING FOR THE KEYS TO THE DISARMAMENT/ARMS
CONTROL PROBLEM 87
WILLIAM EPSTEIN

MONITORING THE WORLD'S LARGEST BIOMASS—
BEFORE THE CANAL COMES THROUGH 93
ROBERT MURRAY WATSON

THE WHIRLYGIG—SAILING TO A NEW SPEED
RECORD? 97
NICHOLAS SHEPPARD

AIR POLLUTION—STOPPED AT THE SUBJECT, IF NOT
AT THE SOURCE? 103
VIRGILIO ANDRÉS OLIVERA

THE SEA CUCUMBER—A NEW FOOD 109
LIONEL T. PENGSON

♛ RECYCLING OPHTHALMIC EQUIPMENT 113
PHILLIP HARRIS HENDRICKSON

♛ A DRUG PROBLEM OR A CULTURE PROBLEM? 117
FLORIAN DELTGEN

A NEW GREAT RIVER FOR THE WORLD? 123
ROBERT LIVINGSTON POMEROY

♛ THAT PARAPLEGICS MAY WALK AGAIN 129
GEORGES MARCEL ANDRÉ DELAMARE

♛ DID THE AZTECS USE WOOD FOR BONE IMPLANT MATERIAL—AND CAN WE? 133
HERBERT KRISTEN

COMMUNITY SURVIVAL IN THE ARCTIC—LESSONS FROM PEOPLE AND ANIMALS 137
JOHN CARROLL FENTRESS

EMERGENCY WATER IN HUGE "SELF-ROLLING" BAGS 143
FRANCISCO ALCALDE PECERO

♛ RETURNING TO THE OPEN SEAS IN MODERN PIROGUES 149
BERNARD ANDRÉ FRANÇOIS DUJARDIN

HAVE WE INVENTED CANCER? 153
MICHAEL RAYMOND ZIMMERMAN

♛ ENGINEERING THE BALANCE BETWEEN MAN, LAND, AND WATER 159
JAROSLAV BALEK

♛ EVALUATING THE ROLE OF PREVENTIVE MEDICINE IN AFRICA 163
JEAN-RAYMOND MISSONGO

DOCUMENTING 181 AMPHIBIAN AND REPTILE SPECIES 169
NATHAN W. COHEN

♛ THE VANISHED VILLAGERS OF AFGHANISTAN—A TWENTIETH-CENTURY ENIGMA 177
HIROSHI FUJII

HUMAN COMMUNICATION WITH GORILLAS 181
FRANCINE GRACE PENELOPE PATTERSON

GETTING PROTEIN TO THE THIRD WORLD VIA SELF-HELP 187
ARUNA FERNANDO

BATTLE CHESS 193
VASUDEVAN MOHANLAL

THE LIGHTWRITER 197
TOBY CHURCHILL

TURNING WATER HYACINTHS INTO AN OPPORTUNITY CROP 201
GODOFREDO G. MONSOD, JR.

THE SEARCH FOR LEONARDO DA VINCI'S "THE BATTLE OF ALGHIARI" 209
JOHN FREDRICH ASMUS

OPERATION JUNGLE SOUNDS 215
GUSTAAF HERMAN TEEUWEN

ACROSS THE SEA IN A SKIN BOAT 219
FERNANDO ALONSO ROMERO

OCEANS OF BEETS 223
DAVID LINDSEY MILNE

TRACKING THE GREAT WHALES 227
ROBERT M. GOODMAN

TURNING WASTE HEAT DIRECTLY INTO ELECTRICAL ENERGY 233
MING LUN YU

SKI BINDINGS WITHOUT BOOTS 239
ANTONIO FAULIN

SEARCHING FOR EXTRATERRESTRIAL INTELLIGENCE 245
RICHARD M. ARNOLD

SELF-HELP HOUSING 249
JULIAN CECIL BOCKS

"SHAKE-PROOF" CAMERAS, BINOCULARS, AND TELESCOPES   253
ADRIAN ANTHONY CECIL MARCH

SAVING THE ABYSSINIAN WOLF   259
KENNETH LEE MARTEN

THE GOLD HAS BEEN THERE FOR 270 YEARS, AND NOW . . .   265
CHARLES PLACIDE GUICHARD

ELEPHANT RANGES, WHERE PEOPLE AND ANIMALS COEXIST   269
ROBERT CHARLES DACRÉS OLIVIER

A SOLAR/WIND WATER PUMP   275
MITCHELL HARRIS ZUCKER

THE MERMAIDS OF THE AMAZON   279
HENDRIK NICOLAUS HOECK

THE FOOLPROOF CAMERA   283
THOMAS KEITH GAYLORD

RECORDING A CULTURE DOOMED TO DISAPPEAR: KALAHARI BUSHMEN   287
CARLOS PEDRO NOLASCO VALIENTE NOAILLES

THE KOMODO DRAGON EXPEDITION   293
BARRY JOHN BROOKE

A WAY TO STOP THE ANNUAL SLAUGHTER OF 200,000 PORPOISES   301
DANIEL A. SHEPARD

LOW-COST ZINC OXIDE FROM RECYCLING WASTES   307
UMA PRASAD MAHAPATRA

CHANGING A POPULATION'S DIET TO PREVENT MALNUTRITION   311
JORGE RAFAEL RESTANIO

A SHOE THAT CAN HELP MILLIONS OF PEOPLE TO WALK AGAIN   315
GEORGE WILLIAM HALL CLARKSON

IDENTIFYING THE CANCER HAZARDS IN OUR
ENVIRONMENT   319
IMRE FEDORCSAK

SOLVING CONSTRUCTION PROBLEMS IN
LOW-TECHNOLOGY AREAS   321
OLIVER MATHESON BULLEY

SOME BATS ARE DEADLY, SOME ARE NOT—BUT HOW
TO KNOW?   325
REXFORD D. LORD

IS IT ATLANTIS, OR SOMETHING ALTOGETHER
DIFFERENT?   329
DAVID DANIEL ZINK

MAKING LABORATORY EQUIPMENT FOR
SCHOOLS   333
LEELARATNE SENANAYAKE

ENTERPRISES IN BRIEF

PROJECT BAOBAB: A SYMBOL FOR CONSERVATION IN
AFRICA   337

A NONPROFIT HOSPITAL FOR SOUTH TAIWAN
MOUNTAIN ABORIGINES   337

THE SCOTTISH TIERRA DEL FUEGO EXPEDITION   338

THE SEARCH FOR THE ANCIENT SEA PEOPLE OF THE
MEDITERRANEAN   338

A NEW HANDLOOM FOR HOBBY OR INCOME   338

ARID-LAND WATER TRANSPORT WITH SOLAR
ENERGY   339

A COMMUNITY-OWNED POWER AND FERTILIZER
PLANT   339

A WAY TO END OVERBOARD DROWNINGS   339

ARRESTING THE TILT IN THE TOWER OF PISA   340

FLYING THE ANDEAN WAVE—1700 KILOMETERS BY
GLIDER   340

THE ECOLOGY AND AQUACULTURE OF THE SPINY LOBSTER   340

"REDUCE THE SIZE OF HUMAN BEINGS BY TWO-THIRDS . . ."   341

THE LONGEST SOUTH-TO-NORTH WALK   341

NERITICA—MANNED UNDERWATER RESEARCH STATION IN THE RED SEA   341

DID THE INCAS USE SOLAR ENERGY LONG BEFORE US?   342

LANGUAGE BEHAVIOR DURING DREAMS AND OTHER SLEEP PERIODS   342

INDIVIDUAL REMINDERS TO CONSERVE WATER   342

THE REMARKABLE MAYYA FLEXILINER DRAWING INSTRUMENT   343

CAPTIVE BREEDING OF LION MARMOSETS   343

HIGH-TECHNOLOGY CONTROL OF DIABETES MELLITUS   343

CAVE RESEARCH AND DEPTH RECORD ATTEMPT IN IRAN   344

IS THE COMMON MARKET REALLY SOMETHING NEW?   344

COULD IT BE PERPETUAL MOTION?   344

THE FUTURE PORT OF BUENOS AIRES   345

TRACKING THE WHOOPING CRANE VIA VOICE PRINTER   345

NEAR ELIMINATION OF BLOOD LOSS IN SURGICAL OPERATIONS   345

A WORKING FLYING SAUCER   346

SEARCHING FOR A LEGENDARY GOLD SHIP   346

A NEW STRUCTURAL FORM FOR ARCHITECTURE— "FOLDINGS"   346

CONSTRUCTING PERFECT VIOLINS   347

THE VICUS CULTURE—AN UNKNOWN CROSSROAD
BETWEEN NORTH AND SOUTH   347

AMPLIFYING THE HORSEPOWER OF A LOW-POWER
SOURCE   347

IMPROVED METHODS FOR THE AQUACULTURE OF THE
RIVER LOBSTER (PRAWN)   348

FLYING EYE SURGEONS HELP THE BLIND TO SEE
AGAIN   348

PROTECTING THE CORAL REEFS   348

MINIATURIZED WASTE RECYLING SYSTEMS FOR THE
THIRD WORLD   349

A LOW-NOISE, CLEAN-EXHAUST AIRPLANE ENGINE   349

PRESERVING THE FOLK MUSIC AND DANCES OF
NORTHERN GREECE   349

SEARCHING FOR THE WORLD'S RAREST PRIMATE   350

THE EARLY DIAGNOSIS OF MALARIA AND
TUBERCULOSIS AT THE COMMUNITY LEVEL   350

EXPLORING INCAN AND MAYAN TRAVELS VIA WEAVING
MOTIFS   350

BY DOG SLED ACROSS THE TOP OF THE WORLD   351

A PERSONAL, PORTABLE MONITORING DEVICE FOR
DIVERS   351

MOBILE SOLAR INFORMATION CENTERS FOR
AUSTRALIA   351

CHECKING ASTROLOGICAL AND COSMOBIOLOGICAL
THEORIES BY COMPUTER   352

HIGH-ALTITUDE FREEFALL PARACHUTE DESCENTS—
RESEARCH AND RECORDS   352

WORLD BUTTERFLY HEADQUARTERS   352

SOUTH AMERICAN MAPS FOR INTERNATIONAL
EXPLORERS   353

ELECTRICAL ENERGY FROM DEEP-SEA CURRENTS   353

ELIMINATION OF HEAVY OIL BY MICROBES   353

HUMAN ENDEAVOR AGAINST THE ODDS   354

THE SHARED-ENTITY CONCEPT: A NEW LOOK AT LIGHT AND OPTICS   354

THE EFFECTS OF HORMONE LEVELS ON THE EFFICIENCY OF NITROGEN FIXATION IN PLANTS   354

A WALK ACROSS EURASIA   355

USING WASTE PRODUCTS TO REFOREST A COUNTRY   355

WATER QUALITY IN THE FOX RIVER VALLEY   355

"BLISSYMBOLS" FOR SAVING LIVES ON THE ROAD   356

THE SQUARING-CIRCLE SPIRAL GRAPH   356

EXPLORING UNDERGROUND SPACE WITH RADAR   356

THE HANDCRAFTED PAPER PROJECT   357

HOMO SAPIENS IN THE WESTERN HEMISPHERE FOR 250,000 YEARS?   357

THE ATLANTIC DRIFT PROJECT   357

SOLAR ENERGY TO POWER AN ELECTRIC VEHICLE   358

ATTEMPTING TO SAVE THE BARBARY LEOPARD   358

SOLO DIRIGIBLE FLIGHT OVER THE ATLANTIC   358

NAME INDEX   359

FOREWORD

This volume celebrates the human nature of the enterprise that made the world we live in today and will change the world again before tomorrow. Here are schemes to feed the two billion who will join us here on Earth in the next two decades; to seek adventure in the sky, at sea, under water, and underground; to save endangered species of plants and animals, including the most endangered species of all, man himself; to put us on speaking terms with our closest relative, the gorilla, and with the orca whose head holds a brain six times bigger than ours; to recover energy at the downhill limits of the Second Law; to bring into the common memory the millennia of human experience recorded in the cave paintings that await discovery in the depths of the Eurasian continent. Each of these enterprises has two partners. One is the entrant. The other is the reader, who is challenged to share the risk and the reward, in imagination, at least, and in action if there is the will to match.

The big machines and the huge organizations of industrial civilization conceal the identity of their true authors. That they seem to run automatically provides an ironic measure of their perfection and success. We need to be reminded that every industrial process is a laboratory experiment that has been scaled up and made to go over and over again or to go continuously. To make it go the first time requires imagination, perception, intelligence, determination, and courage. Such qualitites are found, by definition, in individual human beings, not in machines and not even in organizations. The display of these qualities by the people in this book puts us in close touch with the spirit of enterprise.

That spirit is our distinguishing biological heritage. It first made its presence known in nature two or three million years ago in the behavior of the tool-making hominids that started our differentiation from the rest of the primate order. The record shows, therefore, that the enterprise that went into the shaping of tools shaped also the hand of man and the great hemispheres of his brain. Whether for the solemn purpose of survival or for fun, the ventures proposed in these pages declare the essence of what makes us human.

New York
June 1978

Gerard Piel
Publisher, *Scientific American*

PREFACE

It is a distinct pleasure to introduce the individuals and ideas appearing in this book. They share the experience of having been among the more than 3000 international applicants in "The Rolex Awards for Enterprise." You will meet the five Rolex Laureates, 26 Honorable Mention winners, and many others; their goals and efforts are, I believe, fine examples of the remarkable diversity in the human spirit of enterprise.

Our intention in presenting these people is to encourage them further by helping their ideas to reach a larger audience than they might achieve on their own. Additionally, I hope that this book will contribute to a broadening of support for the spirit of enterprise around the world. In our view, this spirit is an essential ingredient in the world's myriad searches for ways to improve the quality of life.

We at Montres Rolex S.A. in Geneva have always prized that spirit greatly. Since the early days of this century, it has been a particularly important part of our own enterprise in the world of fine watchmaking. As a result, today we can take pride in several watchmaking innovations achieved by Rolex. So in 1976, we came to the 50th anniversary of a "first" that occupies a special niche in our traditions—the invention and patent of the Rolex Oyster Case. Acclaimed in 1926 as the world's first waterproof watch, the Rolex Oyster has housed our many developments since. Given the significance of the Rolex Oyster to us, its 50th anniversary seemed an appropriate occasion to share and celebrate what we have long called "The Spirit of Enterprise."

By inaugurating "The Rolex Awards for Enterprise," we wished to pay tribute and give tangible support to a human characteristic that we value. Given the gratifying response to "The Rolex Awards for Enterprise," we were pleased to go beyond our original plan to encourage enterprising individuals—broadening the awards to include the 26 Honorable Mention winners and arranging for publication of their ideas.

It was my pleasure to announce at the Awards Presentation Ceremony, held on March 2, 1978 in Geneva, that we would present "The Rolex Awards for Enterprise" again in early 1981. In this way we will again be able to foster the spirit of enterprise around the world. It is my sincere hope that this book will help to challenge and stimulate your own spirit of enterprise and inspire you to undertake an endeavor of your own choosing.

Geneva André J. Heiniger
June 1978 *Managing Director*
Montres Rolex S.A.

INTRODUCTION

You are about to encounter an unusual group of people. Because you meet them and their ideas as you read through the book, introductions need not be made now. Yet some background on this group perhaps will help you to appreciate their special talents.

These 131 individuals total about 1% of the slightly more than 12,000 persons from around the globe who wrote to Montres Rolex S.A. of Geneva, Switzerland for applications to "The Rolex Awards for Enterprise." The Awards, offered internationally by Rolex, had been designed to encourage enterprising people and their endeavors in three broad categories: Applied Science and Invention, Exploration and Discovery, and the Environment.

There were to be five Rolex Laureates chosen, each of whom would be invited to Geneva to receive a prize of 50,000 Swiss francs and a gold Rolex Chronometer. To administer "The Rolex Awards for Enterprise," a separate organization was established: The Secretariat, The Rolex Awards for Enterprise, P.O. Box 178, 1211 Geneva 26, Switzerland. (It remains in existence, as will be noted below.)

To adjudicate the awards, a selection committee of internationally known experts was chosen. They were Professor Margaret Burbidge of the University of California, San Diego (and former director of the Greenwich Observatory); Professor Derek A. Jackson, who teaches at the National Scientific Research Center in France; Mr. Luis Marden, former chief of the foreign editorial staff of *National Geographic Magazine;* Professor Jacques Piccard, the well-known oceanologist; and Professor Olivier Reverdin, president of the Swiss National Council for Scientific Research and former president of the Council of Europe. Presiding over the selection committee was Mr. André Heiniger, chief executive and managing director of Montres Rolex S.A.

Requests for applications began to pour into the Secretariat in late 1976, following announcement of the Awards. Each potential entrant was forwarded a detailed, 15-page application form that requested personal information and qualifications, a description of the enterprise project, the names of references who could provide further information if required, and so on. More than 3000 people, representing 88 countries, completed the applications and returned them to Geneva before the March 31, 1977 deadline.

The response was larger than had been anticipated. When the Secretariat made this fact known to Montres Rolex S.A., the company reacted with two decisions. One decision was to extend official recognition beyond that originally limited to the five Laureates. This was accomplished by awarding an additional 26 Honorable Mention winners with a gold-and-steel Rolex Chronometer. The other decision was to provide the Laureates, the Honorable Mention winners, and a selection of other entrants the further support and credit that they might receive through their presentation in this book.

What makes these people an unusual group? Perhaps the best answer is that they comprise a rare collection of individuals all possessed of drive, enthusiasm, excitement, and—simply put—enterprise, as you will see. Apart from the brief introductions and minor editing, you will be reading their own words, from their descriptions of their own projects on the application forms. Collectively they represent a high degree of human ingenuity.

If, as you explore their worlds with them, you find that your own spirit of enterprise is fired or rekindled, let them know. They will welcome your interest, and you will be cementing your own commitment to goals you seek to achieve.

To help strengthen that commitment, in late 1979 the Secretariat will once again have applications available, this time for "The Rolex Awards for Enterprise" that will be given early 1981.

Paris   Gregory B. Stone
June 1978   *Editor*

IN THE SPIRIT OF ENTERPRISE
FROM THE ROLEX AWARDS

OVER THE SOUTH POLE BY BALLOON

To be "the first to . . . " is a recurring theme in the world of exploration and discovery, a challenge that requires ever more inventiveness as our globe yields its secret places to determined adventurers. Too easily forgotten are the relatively recent achievements of some explorers and the arduous efforts required in earlier "man-against-nature" struggles.

Antarctica is a part of the world that not only has continued to capture the imagination of daring and questing people but that still offers a wide range of "firsts" to those who possess the drive, skill, and spirit of enterprise to strive for them. The goal of this project is to be the first to cross the South Pole by balloon, an attempt fraught with very special dangers.

ROLAND READING PARSONS
Honorable Mention, Rolex Awards for Enterprise
Officers' Mess
RNZAF Base Ohakea
Private Bag, Palmerston North
New Zealand
Born April 2, 1941. British.

Squadron Leader Roland R. Parsons, as commanding officer of the Administrative Squadron, RNZAF Base Ohakea, is responsible for administering the accounting, secretarial, medical, personnel, and other tasks of the base. World prizewinner in the economics examinations of the Chartered Institute of Secretaries at London School of Economics in 1961, Parsons went on to further studies in military management and management analysis and administration before being elected a member of the New Zealand Institute of Management in 1973. He is also a fellow of the Royal Geographical Society, London, and an associate member of the Royal Aeronautical Society.

The expedition will be carried out within the geographic limitations of the Ross Dependency (New Zealand), Antarctica, and will have four major objectives.

1. To reenact the pioneering achievement in aviation history of the first Antarctic balloon flight made by Captain Robert Falcon Scott, from near Hutt Point, Erebus Bay, on February 4, 1902.

2. To trace, by low-level balloon flight, the final routes of Amundsen and Scott over the Polar Platcau to the South Pole and in so doing accomplish the first transit of the pole by balloon.

3. To make the first balloon exploration flight over the Trans-Antarctic Mountains. The transit over the Royal Society Range will overview the route Scott, by sledge and without dogs, took from McMurdo Sound to the Polar Plateau in 1903.

4. To conduct a research field study, throughout the expedition, on polar wind systems. The study will include a full meteorological survey and summary of atmospheric conditions existing at the time.

It is axiomatic that the expedition must be filmed, both from the ground and by air. I have had sufficient experience in the management of a camera crew, and several television and private film units have expressed an interest in filming this activity. However, my intention is that the total operational costs of the expedition, other than the Rolex Award, would be met by a film company, by negotiation.

Details of the Expedition

Scott took his balloon ascent with the intention of obtaining distant views above the McMurdo Ice Shelf. I intend to reenact the flight, tethered, like Scott, to about 700 feet, and through the eye of the cameraman see what Scott saw. I am currently locating copies of the photographs that Shackleton is understood to have taken, and I plan to use them in the film as a source of historic interest. Hutt Point itself is full of interest, and Scott's effects are almost intact today. I want the expedition to be self-contained, utilizing wherever possible the resources and expertise of units in Antarctica. The detailed planning and logistics for the expedition are now being prepared, as any operation of this nature must ultimately have the approval of the Ross Dependency Research Committee and the sanction of the Department of Scientific and Industrial Research (Antarctic Division). The reenactment will be located initially at Scott Base, then move to Hutt Point and return to Scott Base.

Our second objective will center on the Amundsen/Scott South Pole Base (United States). Weather conditions and wind direction will determine the nature of the balloon's track, but we plan a transit of up to 20 miles in a direct line across the pole. From the launch spot in the vicinity of Scott's Last Depot, a low-level flight to and across the pole will be undertaken. On completion of this objective, the expedition will return to Scott Base.

The third objective, the balloon transit of the Trans-Antarctic Mountains, will originate from Scott Base but will be deployed in favorable weather conditions to the upper névé basin of the Blue Glacier or to Koettlitz Glacier, where the balloon would be launched. The transit would take place with the secondary objective of clearing Mount Lister (13,200 feet) and completely crossing the Royal Society Range. A landing would be attempted in the wide open northeastern Skelton Névé. On-the-spot research, however, may require a better alternate route. If wind conditions are ideal, a flight of about 80 miles might be direct from Scott Base right over the Royal Society Range. However, for planning purposes we will restrict the flight to about 30 miles and cross the shortest possible distance. Historically, the Royal Society Range has been a major asset in the exploration of Antarctica. The range is easily seen from either Scott (New Zealand) or McMurdo (United States) Bases, and it is natural that Scott himself chose to make a transit of the Trans-Antarctic Mountains.

Our fourth objective, a meteorological study of the polar wind systems, is a natural corollary to this expedition. In the science of aerostatics, of which the flying of manned balloons is an integral part, a thorough knowledge of atmospheric conditions is essential to the successful completion of the overall mission. The New Zealand Meteorological Service is willing to support the expedition and participate in this study. I am personally hopeful that a specialist observer will be made available for the expedition or that a research student of one of our universities will volunteer for this task. In any event, I will have full weather-reporting assistance from either New Zealand, Australia, or United States sources.

General Factors

Size of the Expedition. The success of the expedition requires that the size of the party be restricted. First the balloon (my own Cameron 77,500 cubic-foot balloon), at the maximal load in Antarctic conditions, will lift only three people. There is no advantage to a larger balloon because the larger the balloon, the more difficult it is to operate in mountainous terrain. My own class of balloon has a high performance capability, is new in design and construction, and has already proven itself in mountains and high-altitude use. The expedition members, therefore, will comprise myself (as pilot), a copilot, a ground operations and logistics officer, a meteorologist, a cameraman, and a second cameraman and sound technician. The latter three appointments will be made immediately prior to the expedition.

Accommodation is critical in Antarctica—another reason the size of the expedition must be restricted. Furthermore, every field party must be experienced in snowcraft, icecraft, and alpine survival techniques. We have that experience.

Air Support. If the expedition is endorsed by the Ross Dependency Research Committee, as I am confident it will be, then air support in

Antarctica will be called upon. The cost of such operations will have to be negotiated with the committee and the complete operation included in the New Zealand Antarctic Research Program and funded by the expedition's resources. In New Zealand, my own experience has shown that, with careful management and one helicopter, all support, deployment, recovery, and filming functions can be accomplished economically with no waste in flying time or crew use.

Reconnaissance. With the Royal New Zealand Air Force (RNZAF) I made a short flight to Antarctica in 1974 and cannot see any technical reason why the expedition should not proceed.

Support. Several organizations have indicated their willingness to support the expedition and assist in the technical performance: The Royal Geographic Society, London; Palmerston North Branch of the New Zealand Division, Royal Aeronautical Society; the Canterbury Museum, Christchurch; New Zealand/Antarctica Association, Christchurch; and the New Zealand Meteorological Services. In addition, I am supported by many individual consultants who specialize in the Antarctic.

Special Techniques. The most difficult aspect of this expedition will be the maintenance of propane gas (the fuel for the hot-air balloon) in subzero temperatures. For flight and safety, the gas will have to be kept at an ambient temperature of 10–15° Celsius. I will be studying this problem with the balloon maker. The Royal Aeronautical Society will assist New Zealand industry in making a completely insulated interior for the balloon basket, for additional protection in cold conditions.

Balloon Modifications. The balloon I fly, "West Wind," will be fully instrumented with radio and survival equipment capable of meeting any possible hazard in Antarctica. We intend to build a survival cocoon within the wicker basket, to protect the occupants in the worst of known conditions and to avoid many in-flight problems of exposure. Additionally, the burner equipment quality will be improved to ensure faultless vaporization with delivery of liquid propane in a colder atmosphere. Here I will call on the experience of Julian Nott (high-altitude balloonist of the United Kingdom), specialist advisers of the American Balloon Federation, and the British Balloon and Airship Club. We have the resources in New Zealand to accomplish any special modifications we require, and, more importantly, the engineering expertise of members of the Royal Aeronautical Society would be most helpful.

Cypripedium calceolus.

SAVE THE ORCHIDS

Mention endangered species, and most people reflexively think of whales, certain birds, and other noteworthy animals, in whose defense various committed groups of ecologists have raised public hue and cry against the ravages of industry and urban encroachment. Not many of us pause to think about the lesser known and less well-defended species that are equally endangered—such as orchids.

Delicate and beautiful, orchids exist in a surprising number of varieties, scattered across a remarkably broad spectrum of living conditions. Possessed of a flaw unique in the vegetable kingdom, their survival depends on an exquisitely precarious symbiotic relationship with surrounding nature. That needed environment is clearly being threatened by people. And it may well be a human, whose passionate concern for the future of orchids is shown here, who will be responsible for saving endangered orchids.

GUALTIERO GIOVANDO
Via Fontana Ampia
12069 Santa Vittoria d'Alba
Prov. Cuneo
Italy
Born May 14, 1947. Italian.

In his work over the past five years as chief ecologist for the firm Idronova, Gualtiero Giovando has dealt mainly with the development of equipment and new techniques for water purification. As an agriculturist at the agricultural universities of Sassari and Turin, he learned the cultivation in vitro of vegetable tissue and the sowing in vitro of orchids. This ecological concern is his passion, and in his project he allows his passion to bloom.

How can we save about 10 types of European orchids from certain extinction, and how can we prevent the possible end of another 60 types? Few people know that there are 97 types of orchids in Europe, divided into 26 groups and distributed between 0 and 2600 meters of altitude. You can find orchids on the Canary Islands of Portugal, in the Ural Mountains, the Caucasus, North Africa, and on up to Scandinavia at the 68° parallel. The northern orchids are more modest than their gaudy tropical relatives, but they have an equally complicated biology: They are the only flora whose seeds lack cotyledons. Because such seeds do not germinate spontaneously, orchids must establish symbiotic relationships with molds in order to survive.

Even if we continued to produce thousands of seeds, only a few will produce, in 4 to 10 years, an adult plant. Flora having such a delicate and complicated biology are destined, in a world so deeply colonized by humans, to die out little by little unless serious measures are taken. The following five species are now dying out: *Malaxis monophylla, Liparis loeseli, Spiranthes romanzoffiana, Cypripedium calceolus,* and *Loroglossum hircinum.* The following nine species are now very uncommon: *Calypso bulbosa, Epipactis microphylla, Nigritella rubra, Herminium monorchis, Orchis saccifera, Orchis saccata, Orchis praetermissa, Ophrys speculum,* and *Ophrys bombiliflora.* For the others, time may also be running out.

Forgive me if I speak of myself; I do not do it for presumption, but to make myself better understood. I do not consider ecology as bread, but above all as a passion. I dedicate my leisure time to the European orchids, and, occasionally helped by some keen friends, I have achieved the following things: I have set up a technique for orchid germination in vitro under sterile conditions; three years ago, I founded the International Association for European Orchids; and with many friends and collaborators, I intensively explored well-known places in Europe to see if the native orchids still existed, and in what condition.

By legally forbidding the picking of orchids, all European countries acknowledge the problem. But the legal defense is only passive, not active. By itself, it is not sufficient to stop the progressive, slow extinction of these flora.

I make the following proposals: First, I wish to create a botanical garden for repopulation and acclimation, where the most typical environment is reproduced for the European orchid (this is possible in Italy). The garden would be filled with all types of the transplantable European orchids and thus would be the only garden in the world specially constructed for European orchids where the general public could get acquainted with these flora.

Nine environments would be built: a clearing and border of a deciduous woodland; a clearing and border of a conifer woodland; a deciduous forest; a conifer forest; a thin clay meadow; a stony meadow; a marshland; a moist sand area; and a peatmoss area. A minimum area of 4 hectares (40,000 m^2) is necessary if the preceding habitations are to be effective, that is, to keep a certain autonomy relative to the others.

I am willing to spend all my savings, plus savings of other sympathizers, gathered by me, to buy the meadow. However, we will need additional financial help. We need about 15,000 Swiss francs to prepare the meadow and build all the habitats.

Second, I wish to found a seed laboratory, for orchid reproduction in vitro and to attempt reintroduction of plants in proper environments where they have reached the point of virtual extinction. To do this, I shall use aseptic, in vitro, sowing techniques using symbionts and asymbionts, cultivating meristem tissue in vitro to obtain clonal propagation and cultivating tissue produced by inflorescence.

Third, I wish to complete my census of the European orchids and to report on their distribution and conditions of prosperity. This is the most expensive work, owing to the zone's wideness and the time required for exploration.

Fourth, I should like to publish a bulletin that would appear regularly, to acquaint subscribers and hobbyists with the situation and to help publicize the cause of the European orchids.

Many people say that poetry has died in this century, but I do not think so. Even if there exist more urgent problems for humanity, such as the solution of energy problems or of pollution, I think that there is still time to think of more modest things, such as flowers.

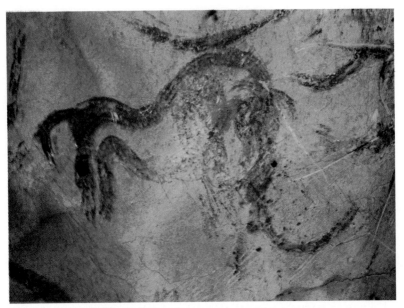

This horse, on a wall in the Baume-Latrone caves, is 1.1 meters long.

THE WORLD'S LARGEST AND OLDEST ART COLLECTION

Perhaps we will never know just what it is about the spirit of enterprise that makes one individual able to move from idea to action, from thoughts that others may have had to the accomplishment that becomes uniquely one's own. Whatever that critical, magical ingredient may be, it exists in large measure in this Rolex Laureate.

Most of us have heard of the ancient, Paleolithic cave paintings at Lascaux and have marveled or mused over these 40,000-year-old messages from our early forebears. Some may have paused over the greater wonder: Forty thousand *years!* (Does anyone dare to imagine anything of one's own making being viewed by descendants 40,000 years from now?) Who were these people of Lascaux, whose ancient civilization makes that of the Egyptian pharoahs look contemporary by comparison? And was Lascaux a fluke, a microcosmic and atypical bubble of advanced culture in a world of "not quite humans"?

The answer is no, it was not a fluke. The 40,000-year-old paintings can be found in caves across Europe, 156 of them by this Rolex Laureate's count. They constitute a collection of art unrivaled anywhere in the world, and recording them—in photograph, map, and diagram form—is the enterprising project of this thoughtful and energetic man.

LUC JEAN-FRANÇOIS DEBECKER
Rolex Laureate, Rolex Awards for Enterprise
5 B Chemic Golette
1217 Meyrin
Geneve
Switzerland
Born July 7, 1934. Belgian.

A surveyor whose work has taken him through most of Europe and into North Africa, Luc Debecker first went "underground" in 1950, when he and his wife were invited to join friends on a weekend speleological foray. That experience sparked a passionate hobby that occupies major portions of the Debeckers' spare time. In 1958, Luc Debecker saw his first Paleolithic cave painting and became fascinated with these 40,000-year-old works of art. By 1968, what had been a healthy interest in the ancient remnants was transformed into a major enterprise. Having searched in vain for adequate reference materials on cave paintings, Luc set out to document and map all the European cave paintings, a staggering undertaking requiring visits to at least 156 caves where the paintings are known to exist.

Luc Debecker (and his family: Mrs. Debecker, two children, and the family dog take part in expeditions to the caves) has probably seen more Paleolithic cave paintings than anyone else in the world. He advances strong reasons for compiling a record of them, and it is likely that the results of his project will lead to new and important understandings of our past.

Thanks to excavations of prehistoric works, the material conditions of our ancestors' lives are quite clear to us. The main obscure point that remains concerns the mentality, beliefs, and knowledge of prehistoric humans.

The most ancient artistic testimony to the vast human experience are the works of art preserved inside caves. These paintings are 40,000-year-old relics. The whole world knows and admires the masterpieces in the sanctuary caves of Lascaux and Altamira. Unfortunately, however, most of the paintings in the 156 known decorated caves remain unknown to the public.

The problems of studying these ancient works of art are almost insurmountable. Whether one wishes to study their chronology or to establish by comparison a complete synthesis of them, the means of accomplishment are severely restricted.

Two works on the subject, Abbé Breuil's *400 Siècles d'Art Pariétal* and Leroy-Gourhan's *L'Art Occidental Préhistorique,* although masterpieces in the literature, are incomplete and insufficiently detailed. Very often, the decorated caves simply have not been described, or have been inadequately described, either because elements that do not support the author's thesis have been omitted or because further discoveries have not been published. Even when published, they are hardly accessible except in rare, specialized, or local libraries. The searchers who undertake to comprehend the mentality of our most antique civilization are faced with enormous blanks.

André Malraux taught us: "L'art est le reflet le plus fidéle d'un période humain" ("Art is the most faithful reflection of a human period")—but how can we understand this prehistoric civilization if we lack a complete inventory of these works, miraculously preserved during millennia? What are the main reasons preventing the realization of such an inventory?

The first obstacle is the geographic distribution of the caves, dispersed in many countries and areas. In France, there is a first zone in the center (mainly in the Lot and the Dordogne), a second zone in the Mediterranean south, and a third zone in the Pyrenees and Bordeaux areas. In Spain, there is the area of Cantabrique and the Asturies in the Pyrenees, a central zone, and the Mediterranean coast in the south. In Italy, there are some caves in the south and in the Sicilian islands.

Second, in these large areas, the exact geographic locations of some decorated caves (those of a less spectacular artistic value) are hard to find. Well known only to local archaeologists or speleologists, they remain unknown to our automobile-based society, lost in places where only walkers can find them. Then, too, it is often necessary to camp at the sites, as board and lodging may not be available nearby.

Third, there are numerous difficulties in gaining access to the cave entrances; some of them require a long approach, made arduous by the occasional need for mountain climbing and walking.

Fourth, visiting some caves can be considered only by people who are trained in speleological techniques and properly equipped. The Montespan cave is a hydraulic bore 1500 meters long, which requires 4–6 hours exploring, mainly under river water that has an approximate temperature of 6–8° Celsius. Both upstream and downstream accesses are barred by wetting or siphoning vaults. Prehistoric engravings and modelings are dispersed all along the journey.

The Etcheberriko cave is situated in the mountains, many hours from the nearest village. To reach the paintings, visitors must cross two underground lakes, using first pneumatic boats and then rope ladders to scale a wall 40 meters high. Neither of these caves is an exception. Many other caves could be added: Baume, Latrone, Cullalvera, le Colombier, and so on.

Fifth, making topographic drawings of the paintings also requires specialized material and professional skills. The use of instruments (level and theodolite) necessary for angular and altitude measurements becomes difficult in a dark, damp environment.

Sixth, similar difficulties hold true for cameras, essentially those of transporting fragile machinery in a hostile environment.

Seventh, although classified as historical monuments, not all of the decorated caves in France are the property of the state, museums, or universities. Most often, they are private property, into which entry is forbidden by an owner who hasn't understood the artistic interest of his property except in return for money.

Details of the Project

The proposed project consists of preparing a report for each decorated cave known to date, listing:

1. The geographical location of the entrance

2. The name of the owner or of the person responsible for permitting visits to the cave

3. A description of the difficulties and materials necessary in visiting and recording the art works

4. A concise description of the works

5. A map placing the works in situ

6. Color reproductions

7. A bibliography of the most complete publications concerning that cave.

The publication of such a document would give prehistorians, art historians, or interested amateurs a reference on which they could rely. It would make available to all what is now in the hands of only a privileged few. It would provide the most complete inventory to date and would allow a comparison of ideas from which will certainly emerge new and original theses about our ancestors' mentality. And last, although probably most important, it will catalogue all caves in danger of quickly vanishing. The numerous contemporary graffiti engraved on the walls already spoil the prehistorical paintings and engravings. Each year, a drawing vanishes because of candle soot and acetylene lamps that imprudent visitors have brought too near to the walls. Frost cracks the rocks into small plates, and moss, lichen, and algae grow in these cracks and contribute to the destruction of works of art situated in the cave entries. Sadly, our own interest in humanity may be the greatest danger to these ancient works of art. If we do not preserve the caves now, over the next 40 years we are likely to destroy these relics from 40,000 years ago.

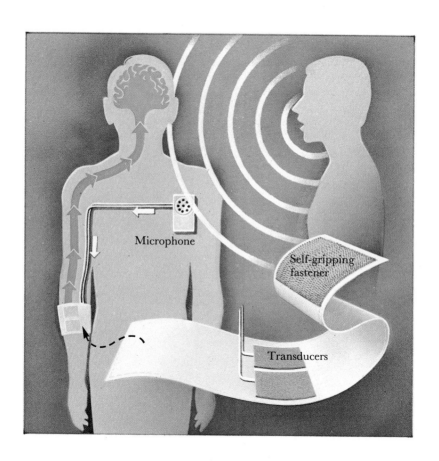

HEARING THROUGH THE SKIN

In the silent world of the totally deaf, life is very different than it is for those of us who possess the gift of even partial hearing. The infant born deaf is instantly disadvantaged, as he or she does not—cannot—know the immensely important human capability of hearing.

By melding present-day technology with the biologist's understanding of our sensory functions, a solution may now be available that could bestow a priceless gift on those who might otherwise never hear the speech of a fellow human, or music, or the sound of raindrops.

DAVID FRANKLIN
9 Preston Road
Somerville, Massachusetts 02143
United States of America
Born August 8, 1932. American.

After taking his B.A. in physics at New York University in 1956, David Franklin became affiliated with the US Naval Applied Science Laboratory as a staff engineer. He subsequently became head of the Electronic Warfare Group, during which time he did graduate work in electronic engineering and physics. His assignment as resident engineer at MIT led to his joining Draper Laboratories, where he became involved in the "Deep Submergence Rescue Vehicle Program Interface Design" and, later, in studying methods for exciting the vibrotactile sense. He has sustained a 15-year interest in tactile communication for the deaf that contributed greatly to the project he outlines here.

The prosthetic ear uses the sensory channel of the skin (tactile sense) to present auditory data to a deaf individual. The device consists of the following: a microphone to receive auditory signals; an amplifying section to bring signals to suitable processing levels; a processing section that compresses the linear frequency of the input, reducing the band width from approximately 3000 Hz to 300 Hz, shifted so that it occupies the band 150–450 Hz; a driver section with two filters designed to divide the output into two unequal signals; and two vibrotactile skin excitors, each driven by one of the filters. The device is inexpensive, small, continuously wearable, and cosmetically acceptable.

Attempts to substitute the tactile channel for the auditory channel are not new. The term *sensory plasticity* has been coined to describe the remarkable adaptability of the human central nervous system in using data meant for one sense when presented to another. All recent experiments known to me, however, have been limited and not deliberately directed to an end product such as a prosthetic ear. My effort differs from past experiments in that it is specifically directed toward creating a prosthetic device, and thus wearability and cosmetic acceptability are serious constraints. Also, I have attempted to stay as close as possible to the processing scheme provided by nature in the form of the basilar membrane in the ear. In particular, linear frequency compression allows this system to present a scaled set of signals to the skin that retain the character of their frequency much as expressed by the inner ear. (Retaining frequency in natural form permits perception of traveling waves.) Stated otherwise, the entire band of frequencies 300–3300 Hz is compressed and shifted so that it lies between 150–450 Hz. This has not been the case in other experiments known to me, although in one some frequency data were retained, with excellent results.

Detailed Description

Figure 1 is a block diagram of the proposed device.

Microphone. The microphone is a standard unit of the type used in hearing aids.

Amplifier and Envelope Detector. This section is conventional, using low-noise-integrated circuits. The envelope detector, consisting of a low pass filter and detector, generates an amplitude control signal at the output. This is necessary because processing in the linear frequency compressor results in the loss of envelope information. This section will include a circuit that keeps signals at suitable processing levels.

Linear Frequency Compressor. The circuit to perform this function is in Figure 2. This method follows in essential detail a technique described by J. Baguet and P. Marcou. The system operates as follows: The original signal, $f(t)$, is up-converted through a single sideband (SSB) modulator to become the signal $f'(t)$. This signal, occupying the band 100,300 Hz to 103,300 Hz, is hard limited, passed through a digital counter functioning as a divide-by-10 network, bandpass filtered and mixed with a 10,120-Hz local oscillator (LO) signal. The resulting output is low pass filtered, and the remaining signal, $f(t)/10 + 120$ Hz, contains all the original speech information, although compressed in form. That is, it now occupies the band 150–450 Hz.

Driver. The required power capability of the output driver can be extrapolated from available data on the B98 bone conduction unit manufactured by Radio Ear Corp. For these units, threshold drive for

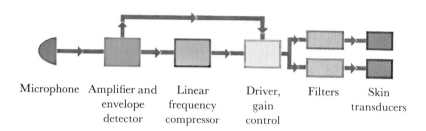

Figure 1.

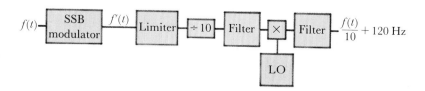

Figure 2.

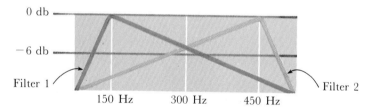

Figure 3.

excitation on the palm is on the order of 10^{-5} watts. If we allow for 60-db difference in required excursion levels in this application and assume excursion increases linearly with voltage, maximum power should be on the order of 10^{-2} watts per transducer. Because two transducers are used, a driver rated at something above 20 milliwatts suffices. Incorporated in this section are a manual gain control for the wearer to use at his or her convenience. Also in this section is the amplitude control element (modulator), which receives a signal from the envelope detector and restores envelope shape to the output signal.

Filters. Two output filters are driven in parallel, each filter having the response characteristics shown in Figure 3. The output of each filter drives one skin transducer.

Transducers. The two skin transducers are of the B98 Radio Ear type. They are mounted on elasticized cloth tape with self-gripping fastening material at either end of the tape. The mounting arrangement for attachment on the forearm is shown on page 16. As demonstrated by Alles and others, this arrangement of transducers gives rise to the perception of a point of stimulation between the two transducers at a position proportional to the relative stimulation intensity of each transducer. Hence, with the filter arrangement shown, the person perceives a smooth mapping of frequency between the two transducers, much as in the basilar membrane.

This constitutes the system. Although the forearm was mentioned as a location for wearing the transducers, other suitable locations include the back, the stomach, or best of all, the palm. The forearm, however, is probably the best compromise from an esthetic and utilitarian point of view. The electronics and microphone will constitute a package about the size of a small cigarette pack and can be worn in a breast pocket or on a cord around the neck.

Evaluation of the device will be accomplished in two steps. The first, a short-term laboratory study, will do pattern recognition tests with both hearing and deaf adults. Design changes will be introduced as necessary. Then units will be evaluated by a clinical group. Discussions have been held with the Tactile Aids Project of the Rehabilitation Engineering Center of the Harvard–MIT Health Sciences and Technical Consortium. This group, which has one or more rudimentary tactile aids under development for clinical evaluation with deaf infants, has indicated interest in evaluating this device. No formal arrangements have yet been made.

Expected duration of the project would be from December 1977 through December 1978.

22

A SMALL, TOUGH HOVERCRAFT FOR BIG, TOUGH RIVERS

In many parts of the world, large and placid rivers have been a boon to humanity, opening communities to others and serving as great roads of communication and trade. In other parts of the world, though, massive rivers do not unite people but keep them apart by denying easy transport.

Drawn by the need for water, the prospect of fishing for food, and some minimal level of transportation possibilities, communities spring up even on inhospitable rivers and manage to thrive. Often these riverbank societies are inaccessible, or nearly so, from the outside world. Seasonal, barely passable roads may link them to "civilization," or there may be no roads at all, travel being limited by mountains, jungles, or other rough terrain. In such situations, emergencies such as serious accidents or illnesses take on added urgency. How does one remove a patient to a place for treatment or bring in necessary medication? In this project, the objective is to find a way to beat the bad rivers, and the first target is high up in faraway Nepal.

MICHAEL EDWIN COLE
Honorable Mention, Rolex Awards for Enterprise
41 Stirling Road
Albrighton, Wolverhampton
West Midlands
England
Born April 10, 1935. British.

Specializing in geography, physical education, and divinity, Michael E. Cole trained as a schoolmaster at King Alfred's College, Winchester (1955–57), and earned the Carnegie College diploma in physical education in 1960. Holding the rank of squadron leader in the Royal Air Force (RAF), he commands the large RAF sports center at Cosford, near Wolverhampton. He is responsible for the resource, initiative, and expedition training of over 1000 apprentices and trainees, as well as for all RAF nordic ski activities.

In my exploration, I plan to use a small and inexpensive four-seat hovercraft on one of Nepal's main rivers, the Kali Gandaki. This fast-flowing, rapids-filled river is a natural highway through an area lacking effective communications.

The HUMP hovercraft has been specifically designed by a missionary engineer for communication and rural health purposes. This hovercraft represents an engineering breakthrough in the use of small craft on fast-flowing rivers. The hovercraft has been renamed the *River Rover,* as it has the potential on rough water that the Land Rover has over rough landscape.

Background to the Proposed Exploration

The high peaks of the Himalayas have now been thoroughly explored, but the mighty river courses of Nepal are less known and are little traveled. For centuries, the only routes through the western areas of Nepal have been narrow trails and high mountain passes. Only two roads have been built in the whole area, tortuous trackways that require large workforces to keep them usable after the ravages of landslides and monsoon rains. Mechanical transport is unknown over nearly all the western area of Nepal.

The fast-flowing rivers are natural highways, but they are largely unnavigable owing to the water speed and frequent rapids. One such river, the Kali Gandaki, flows south from the Mustang on the Tibetan border and then cuts a gap in the Himalayan range between the Annapurna and Dhaulagira Peaks. This fast-flowing river then plunges south for about 25 miles before turning east through the Mahabarat Range. There the river is less turbulent, although use of conventional river craft is still prevented by the fast river flow and numerous, rock-strewn rapids.

This lack of effective communication and transportation is particularly detrimental to the progress of rural health schemes; many patients die on the long treks for medical help. The development of a practical means of transport for these river highways would bring immense benefit to thousands of isolated villagers. A small, lightweight hovercraft has been built to meet this need.

Development of the HUMP (Hovercraft, Utility Multipurpose)

Tim Longley of the Missionary Aviation Fellowship has carried out some highly regarded pioneer work on lightweight hovercraft. He had already designed and built a small hovercraft for use on lakes. The Royal Navy has been so impressed with the design capabilities of the HUMP that the first production model has been produced, for Royal Navy use as a rescue craft, at the Aircraft Yard at Fleetland Gosport. The second HUMP, which is now available, has been earmarked for my use on a possible expedition to the Kali Gandaki River in Nepal. I believe that this maneuverable craft could make medical work possible in hitherto inaccessible areas. This offer to use the *River Rover* represents a unique opportunity to take a fine British invention into challenging expedition circumstances.

Planning Expedition

I led a two-man planning expedition to Nepal during November and December 1975, to explore the feasibility of operating a small hovercraft on the Kali Gandaki River. This included a study of the terrain and of the logistic and diplomatic problems of undertaking a major expedition in the area.

I have taken part in many expeditions for many years, in the spirit of taking on obstacles simply because they are there. In January 1974, for example, I took part in an expedition to bring a water-drilling rig from the United Kingdom to the Wollo Province in the famine area of Ethiopia, in search of life-giving water for the thousands dying of thirst. The situation and our discovery of water provided the ingredients for the most worthwhile expedition I had ever undertaken. The proposed expedition to Nepal should offer a challenging adventure in difficult and wild terrain, as well as the possibility of providing a life-giving means of communication to a pioneer health program.

Purpose of the Expedition

The purpose of this expedition would be to explore the major part of the Kali Gandaki River (over 150 miles), working upriver in the *River Rover*. The hovercraft would be supported by canoes and by an inflated Gemini boat powered by an outboard engine. The nature of the terrain is such that certain areas of the river must be explored on foot.

Each river obstacle would be surveyed and mapped and the performance of the *River Rover* assessed. On the way, four staging posts would be constructed along the river to service the hovercraft and to support the progress of the expedition. These staging posts will be constructed from local timber.

For a distance of 60 miles, the Kali Gandaki River marks the northern border of the Palpa District. The rural health program for this area is hampered by impossible communications. It takes a five-day trek to reach a health post on the river from the only hospital in the whole district, located at Tansen. The English doctor in charge of rural health in the Palpa District hopes that four to six clinics could be set up along the river and be linked by a hovercraft service. This would result in better use of the doctor's and nurses' time and would provide the chance to transport seriously ill patients to the hospital before they die. This project has tremendous humanitarian potential.

Stretches of the upper Kali Gandaki are punctured with cataracts and extremely difficult rapids; these will test the *River Rover* to its limits and will certainly require man-handling the craft around some of these obstacles. As the river rises with the monsoon rains from June until September each year, the ideal time for the expedition would be from September through March, to facilitate studying the river at differing water levels.

EXPERIMENTAL FREE STATES

Given the way the world's "knowledge industry" has exploded over the last 30 years and given the great emphasis on testing industrial systems, products, new drugs and medicines, and so on, it seems odd that we have not, as proposed here, tested our options in the world of sociopolitical behavior. Why not? It seems that each of our political systems has room for improvement, and many people would relish the opportunity to establish a more perfect system.

The enterprising project that follows poses a provocative challenge to the world, one that, if met, might have long-term beneficial results for humanity.

CHRISTIAN HENRIK SNELLMAN
Via Sindicato 43
Palma de Mallorca
Spain
Born August 16, 1930. Finnish.

After completing high school in Helsinki in 1946, Christian H. Snellman went on in 1965–66 to take degrees in political science and history of art at the Umea Branch of the Uppsala University in Sweden. He describes himself as a writer and a journalist.

I propose the creation of miniature states, under the supervision of the United Nations, in different parts of the world, for the development and study of political systems that offer new approaches to communal, socioeconomic, agricultural, environmental, and other aspects of society. These experimental free states would be populated by volunteers from all over the world.

Some of the greatest problems in today's world involve conflicts between established societies and people who want to change them, as well as between groups that have the power and means to reshape the environment and people who do not have the power but do have the desire to conserve and maintain the environment.

New generations see a world gone wrong but are powerless to make more than minimal changes. They are powerless because dissidents generally are in a minority, while the majority consists of people already so involved in, and dependent on, the existing society that they could not change it even if they wanted to.

Until a couple of generations ago, people with the true pioneer spirit, who wanted to start afresh, shaping their own society and environment as they chose, still had a chance to do so, by colonizing new lands. New philosophies and important inventions followed in the wake of such pioneers. Today, the situation is different: The world is getting crowded; people are born in developed societies and environments that allow few alterations; and there are no more "new" lands to settle. Even the most ardently idealistic politicians cannot do much about the conditions they set out to improve.

Such constriction is frustrating. In their bitterness, more and more marginal, dissident groups resort to open war with their societies, which automatically respond with persecution and repression. Thus ideas that at least in theory are quite feasible and correspond to the interests of large groups of citizens are often never tried.

Background

Since World War II, one new society that has been created is Israel. And, in spite of Israel's continuous war with its neighbors, and even taking into account the massive economic aid it has received from the United States, Israel's enormous achievements bear strong testimony to the stimulation of allowing citizens to participate in creating a new society.

In the same epoch, our world has also experienced several revolutions. The subsequent restructurings of the affected countries have sometimes brought improvements from the point of view of their inhabitants, but often at considerable cost of human life. Examples of such revolutions are China, Cuba, and several African states.

Description of the Project

This project, which must remain an abstract concept until the United Nations and at least some nations support it, suggests the creation of social, political, and juridical vacuums. Such vacuums will provide opportunities for fresh starts on territories leased or given by various countries to the United Nations for building alternative societies and environments under UN supervision.

The following development is suggested: First, we need a thorough study of the possibilities of the project, particularly regarding suitable territories and the effect the current international laws would have on the project. Obviously, densely populated countries such as Britain, Holland, Germany, and Japan could spare only very small areas, whereas the situation would be easier for such nations as the USSR, US, and Canada. Even in such relatively small countries as Sweden and Finland, the concentration of people in the cities has left large territories relatively unpopulated.

Second, the project would be suggested to the Secretary General of the United Nations, which would consult its members on the issue.

Third, once sufficient backing was received from at least some of the member countries, the United Nations would create a special experimental free state department, with subdivisions for territory planning, immigration, information and study, finance, nonintervention and peace control, commercial and technical advice, administrative services, and so on.

Fourth, the territory planning department would study the territories offered by the participating nations and would examine their suitability, population capacity, natural resources, and so forth. The existence of natural borders should always be taken into consideration: The ideal solution would be an island. Territories should not be too small, but there should be great flexibility in adapting the project to existing possibilities. The evacuation of the existing population would have to be solved by the host country, and the original inhabitants should be given an option to reenter as free state citizens.

At the fifth stage, the question arises of whether the future citizens of the territory should be selected by their interest in a certain aim or should be allowed to develop policies freely as they go along. If the experimental free states really are to be creative vacuums, as little influence should be exerted as possible. However, it would be valuable if the pioneers of a specific territory had a mutual aim, preferably different from that of the other territories. A humanitarian outlook and a full acceptance of the UN Declaration of Human Rights would be required. A rigid host nation that insisted on a certain program could effectively destroy the freedom of such a state. Yet sponsoring nations might only grant territory on these terms. This question should be settled at this stage between the host nation and the United Nations.

Sixth, the final, bilateral, lease treaty should be drafted in the best interests of both the host country, which should not lose sovereignty over the territory (although in fact leasing it on an idefinite basis), and of the new immigrants, who, once settled, should not have to risk ever having to leave the communities they are creating. Since this project is intended to further international development, the prestige of having provided a territory to this cause might prove to be a main attraction for many nations, especially given the continuous competition for status among the powers.

Seventh, the UN-protected free states would be open to any non-criminal world citizen—regardless of nationality, race, religion, or age—

willing to exchange current citizenship for that of the UN free states and to prove real interest by depositing in the free state bank a sum based on current income. Immigrants would also have to swear allegiance to a simple code of rules, including the UN Declaration of Human Rights and a promise to act only in the interest of the free state and to abstain from all political activities directed against other nations.

The United Nations would publicize the free state projects and activities and would also offer administrative services to the free states until their own governments had been formed. Particularly in the beginning, financing would depend heavily on grants from UN member nations. UN troops would ensure nonintervention in the free states.

I believe this project would offer an important new role for the United Nations and for its participant nations and could prove a new hope for world improvement. It would be an effort of the world to heal itself, and might be a first step toward a truly worldwide United Nations.

FLYING HIGHER THAN EVER BEFORE

In an era of moon landings, orbiting laboratories, and space probes of distant planets in our solar system, precious little seems to remain of the individual's role in pushing back the barriers to flight. We sometimes let slide the memories of the lone strugglers whose beliefs, capabilities, and determination wrenched one triumph after another from a reluctant nature. Today, governments and vast organizations dominate the record books of what we call *flight*.

For many people, spacecraft and other rocket-propelled vehicles belong to another discipline, technologically far removed from real flight. In this flying world, itself heavily dominated by the resources of those self-same governments and organizations, there is a man who believes that *individuals* can still achieve the threshold-breaking successes that have long inspired earthbound imaginations.

DARRYL GEORGE GREENAMYER
P.O. Box 5548
Mission Hills, California 91345
United States of America
Born August 13, 1936. American.

When submitting his application, Darryl G. Greenamyer described himself as "presently employed in the business of contract test pilot assignments for various private aviation companies," an appropriate occupation for someone who had worked for Lockheed Aircraft Company as an engineering test pilot. During his 14 years with Lockheed, he performed a wide variety of test flight activities on several advanced jet aircraft programs, including test flying work on the F-104 Starfighter and the SR-71 and U-2 reconnaissance aircraft. His academic background includes a B.S. in mechanical engineering from the University of Arizona (1961) and graduation from three different professional institutions—the US Air Force Flight Training School (1957), the US Air Force Aerospace Research Pilot School (1964), and the US Navy Fighter Training School (1972).

In 1970, Greenamyer received the Ivan Kinchloe Award for the most significant contribution to experimental flight, given by the Society of Experimental Test Pilots. He shared the award with astronauts Armstrong, Aldrin, and Collins, who were recognized for their first landing on the moon.

I intend to pilot an aircraft to a height above the earth greater than any other man has yet been able to achieve (except, of course, spacecraft and other, similar rocket-propelled vehicles). In so doing, I will establish a new world absolute altitude record for normal, piloted aircraft.

For the past 11 years, I have been involved in a project to explore the boundary of high-altitude flight. This has been a personal effort based on my own desire and conviction of eventual success. I have been told many times over the years that what I am attempting is both futile and impossible, that accomplishments of such scope in aviation are no longer the province of the individual. Friends and experts offered sincere advice that only large government agencies or industry could engage in this type of pioneering because of the cost and technical barriers. Unfortunately, neither government nor industry has seemed motivated or curious enough to accept this challenge. So, although I am not a person of extreme wealth, I continued alone. Besides, I do not like the idea in general that individuals have no place in the realm of flight, no matter how exotic. At present, I have come very close to fully proving my belief.

I needed two things to achieve my objective: a qualified pilot and an airplane of suitable performance. Because my ability to pilot the flight was not a question, all of the effort over the years was spent on constructing the aircraft. In testimony to the negative advice of my friends, I will say that this was a most difficult task. What I needed was a ship capable of speeds in excess of 2.5 times the speed of sound, one over which I could maintain some control in very thin air at 24 miles (38.5 kilometers) altitude. It has taken the very best aircraft of the Soviet Air Force, the MIG 25 Foxbat (E-266 type), to struggle its way to a height of 22.5 miles (36 kilometers) and the current world absolute altitude record. My ship must be better.

The airplane I chose was the Lockheed F-104 Starfighter. This was not exactly the latest model in supersonic jets, having been put into service more than 20 years ago. However, I gained considerable experience flying this aircraft during my test pilot employment with Lockheed, and I am convinced of its basic capability for my special flight. Choosing the F-104 and obtaining an airplane were two different things, however.

Because my project was entirely personal, there was no possibility of obtaining one from the US government on loan or on any other basis. Also, the plane was never extensively employed by the US Air Force; far greater numbers of the F-104 were put into service by foreign air forces. My search began, then, in the metal scrap and salvage yards across the United States. Due to accidents, a small number of F-104 ships had found their way, in piece and part, to such places. Year after year, I pressed my hunt for any type of repairable component. Being a relatively modern supersonic fighter plane, the F-104 is an extremely complex machine. It is made up of an endless number of sophisticated parts, both tiny and large, all assembled under special conditions with exotic techniques. Thus, finding the right parts was only the beginning of my work. It was an agonizingly slow process and often seemed impossible, especially as I tried to do this in my home or in whatever other space I could borrow. Luckily, I had occasional help from friends, engineering specialists familiar with the F-104 and its design, who assisted me with valuable advice.

Gradually, all of the cast-off bits and pieces began to take shape. The bent and torn metal was reformed, the myriad mechanical devices were rebuilt, and, inch by inch, the skeleton of a Starfighter began to appear. Along the way, it was necessary to make numerous modifications as well. I knew the limits of the basic airplane very well and realized that in original configuration it would not allow me to reach the altitude I desired. Introducing experimental devices and modifications further complicated the reassembly process. Feeling myself growing older each day, week, and month made me wonder if I would ever see the end of the whole business. The project was also a constant drain on my financial resources and often would slow or stop as my income dwindled.

It now seems like a century since I first began this quest, but when I look at the sleek and completed ship sitting in its hangar I know it was worth all the trouble. I have taken it into the air, and it flies—something I began to have definite reservations about as the time of the first flight approached. But it does fly, and in a way to which no words can do justice. Only weeks after the maiden flight, and while still making initial tests, I took the Starfighter to a nearby air speed course facility. I wanted to determine if the ship would come through with the required air speed, a level considerably higher than the original F-104 would reach. At low altitude (within 100 meters of the ground), the Starfighter was measured at over 1000 miles an hour (over 1600 kilometers per hour). This was more than I had hoped for. That flight established a new world absolute speed record for any type of aircraft at low altitude. I now know that this ship will fly higher than any other—that it will take me up where the sky is black and starry at mid-day and where finally its engine will cease to operate for lack of air. Up there, I will have to wear a full-pressure spacesuit and have small rockets mounted on the ship's nose to keep on an even keel, for the normal flight controls will have no effect in the ultrathin air. It will be a strange place, and I plan to be the first man to take such an airplane there.

The actual flight is now planned for sometime in the late spring of 1977. This plan is based on the occurrence of favorable upper altitude winds at that time and also on the work still to be done to prepare the high-altitude devices installed on the plane. The flight will probably take place at Edwards Air Force Base, on the Mojave Desert in California. In order to verify my height precisely, I will have to rent the special high-altitude tracking facility located at that base.

The flight profile on the record attempt will be to take off and climb to about 40,000 feet (12,200 meters). At this level, I must accelerate to Mach 2.5 in straight flight. On reaching this speed, which is a critical maximum because of the severe air friction heating of the ship's surface, I will pull the nose up to a climb angle of 60 degrees. As I ascend past 80,000 feet (24,400 meters) altitude, the engine afterburner will blow out and cease to provide any thrust. From this point, I will continue to climb strictly on momentum. At 24 miles (38.5 kilometers) altitude, the ship will have slowed to about 80 miles per hour (128 kilometers per hour) as it reaches the top of its arc of flight. Now the rocket controls must be employed to prevent the airplane from beginning a nose-over-tail tumble and becoming totally uncontrollable at such slow speeds and in such thin air. If all goes as planned, the ship will then recover speed as it dives for the earth. Below 30,000 feet (9150 meters), I will attempt to restart the engine and resume normal flight. The total flight should take less than ten minutes. Since this has never been done before, the time must remain approximate.

What I am doing is my own personal goal; it is something I feel I must do. At the same time, I hope it can serve as an example or a proof of what is really possible. No matter how large or complex we grow in our affairs, individuals can still accomplish and achieve what others may call dreams.

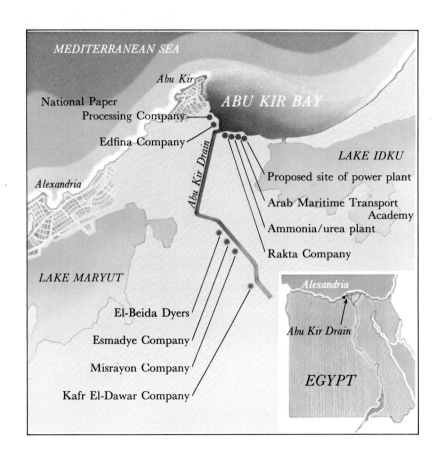

THE ABOU KIR DRAIN — POLLUTION COMES TO DEVELOPING COUNTRIES

Industrialized nations are rapidly losing their monopoly on the evils of industrial pollution as developing countries, intent on moving their economies to higher levels of production, locate large manufacturing operations where available resources and facilities make them most efficient. Near Alexandria, Egypt, manufacturing facilities have already placed an undue burden on a key waterway. Additional construction, planned or under way, will further strain the water resources of the area.

In what could be a classic case history, of value to many other similarly developing areas in the world, a dedicated engineer is trying to assess the situation and find reasonable solutions. Her project seeks not only to rectify the present problems but also to help prevent similar problems in the future.

SAMIA GALAL SAAD
Honorable Mention, Rolex Awards for Enterprise
High Institute of Public Health
Environmental Health Department
165 El-Horreya Avenue
Alexandria
Egypt
Born May 16, 1944. Egyptian.

As assistant professor of sanitary engineering in the Environmental Health Department, Samia G. Saad is responsible for teaching postgraduate students the chemical and the engineering aspects of environmental sanitation. She is also deeply involved in research on industrial pollution control and abatement, as well as with recycling techniques for water conservation in industry.

In 1967, she received her diploma of public health (sanitary chemistry) from the High Institute of Public Health, the University of Alexandria, and then went on to take her master of civil engineering (M.C.E.) at North Carolina State University (NCSU) in 1970. In 1973, she earned her Ph.D. in civil engineering, also from NCSU, before returning to Alexandria to work on the kind of problem she describes here.

In the project, I plan to monitor and submit proposals concerning the pollution created by Kafr El-Dawar, an industrial city, as well as by the pulp and paper factories that dump their wastes into the Abou Kir drain. This drain receives all these untreated wastes, which finally pour into Abou Kir Bay, where the seawater is heavily polluted and hence has sharply declined in value for fishing. Waste treatment methods will be sought to minimize the pollution loads dumped into the canal. Dilution factors, both in the canal and in the bay, must be assessed.

Introduction

Convergent increases in water pollution and water requirements associated with the social and technological developments of the past decade have presented a serious challenge to engineers and others directly concerned with protecting our natural resources and providing water supplies adequate to meet social and industrial demands in our country. Egypt is a developing country where industrialization is making vast progress. The largest industrial complex, at Kafr El-Dawar, near the Alexandria metropolitan area, contains many big factories making textiles (synthetic and natural fibers), dyes and their intermediates, and textile-finishing operations. This complex deposits all its wastes into the Abou Kir Canal, which is used for domestic purposes by the peasants who live on its banks.

The canal also carries drain water from the agricultural areas through which it passes. When it reaches the El-Tarh area, the canal receives raw wastes from two factories producing pulp and paper. Finally, this canal flows into Abou Kir Bay.

All these industries dump their wastes almost without treatment other than simple pH neutralization. As a matter of fact, these wastes have never been analyzed and measured. The impact of all these sources of pollution can be seen in the practically dead stream. Pollution is also evident in the decreased fish catch at Abou Kir Bay, which was once a rich source of fish. And bathers on the shores of the bay complain about the unsightly colored water, which has fibers floating on its surface.

Kafr El-Dawar comprises the following factories:

Kafr El-Dawar Company. Ranks third in the nation for the production of fine cotton textiles and of blends with synthetic fibers. They produce other products, such as medical cotton and sewing threads. This company draws its water from a special branch of the Mahmoudia Canal, which is the main source of drinking water for Alexandria and its environs. The waste water, which is full of starchy sizing, is dumped into an agricultural drain, which then pours into Abou Kir Canal.

El-Beida Dyers. This is one of the three biggest factories for dyeing and finishing textiles in Egypt. They take all of Kafr El-Dawar Company's production and process it for further bleaching, dyeing, and final finishing. They now also have their own plants, which produce woolen fabrics as well as synthetic knitted and woven material. Their wastes comprise a tremendous load of organic polluting substances. These wastes are discharged practically untreated, except for pH adjustment, into the Abou Kir drain.

Misrayon Company. This is the foremost company in the production of viscose rayon from short cellulose fibers obtained from wood. In viscose preparation, strong alkaline soda ash and hydrogen sulfide are applied to imported bleached cellulose sheets. This process yields viscose liquid from which the company produces either rayon fibers by acid bath spinning or cellophane wrap sheets and fibran sheets.

The company also weaves the rayon fibers and applies several dyeing techniques to a variety of rayon products. Their waste is typical of rayon-processing wastes, as well as dye wastes, and it also contains chemicals used as additives and intermediates in this industry. They dump their waste, with no treatment other than neutralization and sedimentation, into an irrigation drain that finally pours into the main Abou Kir drain.

Esmadye Company. Considered the biggest dye company, Esmadye manufactures direct, vat, and mordant dyes. Esmadye imports part of its raw materials, but the rest are manufactured on site. This company has many products, and consequently many wastes are discharged, after neutralization and plain sedimentation, into a branch of the main Abou Kir drain.

The Abou Kir drain carries these wastes about 40 kilometers to El-Tarh Province, close to the shore of the Mediterranean Sea and directly south of Abou Kir Bay. At El-Tarh are three large factories, two for paper processing and the third for food canning. Rakta, one of the paper-processing plants, uses bagasse and rice straw as raw materials for paper production. All paper-processing operations, such as cooking, pulping, bleaching, and paper making, are performed at this company. The National Paper Processing Company depends on used paper, rags, and cardboard to produce carton and wrapping paper. All three companies dump their wastes completely untreated into the Abou Kir drain, just before its final discharge into Abou Kir Bay.

El-Tarh Province is a future industrial expansion area because of its location near the sea and near raw canal water for industrial water intake. Already a new factory is being erected for the production of urea fertilizer. Also, foundations are being built for a new power plant about 20 kilometers away from the fertilizer plant. A new military and naval academy for Arab countries is also in the final stages, about 10 kilometers from the fertilizer plant. Its wastes are intended to be dumped into Abou Kir Bay.

So the situation is becoming critical, and the new factories are now asking what they should do to protect Abou Kir Bay.

Plan of Research

This project requires a period of three years. In the first year, the wastes of all factories, both at Kafr El-Dawar and at El-Tarh, must be surveyed. All the standard chemical analyses of waste waters must be carried out, as well as calculations of pollution loads. Samples along the Abou Kir drain must be taken, as well as samples from Abou Kir Bay, to complete the jigsaw puzzle of pollution impacts on the drain and the bay. Inplant control measures also must be taken into consideration as a useful measure of pollution abatement.

During the second year, mathematical models and pilot simulation studies for economical methods of treatment must be established. A pilot plant, now under construction at the High Institute of Public Health, can be used, with the addition of certain modifications, to treat the variable wastes.

In the last year, we will apply the pilot-scale measures on the industrial scale. We shall make recommendations, for both established and newly constructed companies, for minimizing the pollution of the environment, especially of water bodies.

Project Justification

First, this project will provide environmental ecologists *and* industrialists a deeper insight into how to treat industrial wastes scientifically.

Second, the intended further use of El-Tarh Province as an industrial area requires more emphasis on the bay water, which will be used for cooling purposes by an electric power station and will then be returned to the bay. The bay will also receive the wastes from the new fertilizer plant and the military academy.

Third, this project will give similar factories better ideas about what they can do with their wastes, as they will have a proven example of pollution abatement.

Fourth, the canal will be studied for its water characteristics and its effect on the bay.

Fifth, this project will give graduate students at the High Institute of Public Health a chance to work on real problems of pollution control and acquire a better understanding of environmental and industrial pollution impacts. The project will also help these industries to minimize the size of their pollution problems.

Orcinus orca, *the killer whale.*

DIVING WITH KILLER WHALES

Until recently, few of us could have been blamed for being willing to give a wide berth in waters known to harbor the feared "killer whale." Both ancient folklore and modern media occasionally scare us out of our wits with horror stories about killer whale vengeance, aggression, and the like. For an allegedly rapacious beast, the true stories of calamity are strangely few and far between.

It is beginning to look as if the stories of attack and mayhem concerned a much different kind of whale than we might have imagined—one that reacts in surprisingly human fashion to threats on itself or its family. The more we learn about the killer whales, the more we realize how very close we are to them in development. Even so, it takes more than a little enterprise to consider socializing with these creatures to learn more about them. That, however, is exactly what this entrant intends to do.

MARK CHRISTIAN OVERLAND
1602 Weatherswood Drive, N.W.
Gig Harbor, Washington 98335
United States of America
Born June 28, 1954. American.

As a self-employed researcher and photographer of killer whales, Mark Overland has moved rapidly into a field where too little is known. After taking his B.A. in 1976 from Evergreen State College in Olympia, Washington, where he majored in communications media and cetology and taught photography, he embarked on a highly specialized endeavor. As coordinator of the First International Orca Symposium (held at Evergreen State College in 1976), he was well positioned for the work he does now, which is aided by his being an advanced scuba diver spending many hours underwater with dolphins in aquaria.

Mine is a scientific research project involving the underwater encounter with and complete documentation of human divers and free-roaming killer whales. My purpose is threefold: to provide unique essential footage for a feature-length documentary film, to foster further trust and cooperation between humans and wild killer whales, and to provide scientific data on human–orcan interactions for future ethological research with *Orcinus orca* in a natural setting without artificial controls or motivation.

Our relationship to cetaceans originated in the Stone Age, and the dolphin family (Delphinidae) has attracted great interest, speculation, and admiration. We now recognize that dolphins and porpoises have an advanced intelligence and social life. Of all the dolphins, the killer whale—or orca, as many now prefer to call them—is perhaps the most intriguing species. Worshipped by the Indians, cursed and feared by an ignorant public, the killer whale has been a symbol of terror and cruelty throughout the ages. Only recently have we discovered that for some unexplained reason killer whales are friendly toward us.

Orcas are members of the family Delphinidae and, like all dolphins, they possess highly organized central nervous systems. Surprisingly, the dolphin's neuroencephalization quotient has evolved to a level shared only by humans. Furthermore, certain portions of the dolphin brain are more developed than comparative regions in the human brain. Much of this additional brain development is in the highly convoluted cerebral cortex—an expansion of that part of the brain serving such higher intellectual functions as language, perception, and thought. In this regard, orcas, with an approximate brain weight of 6000 grams (human brains weigh around 1500 grams), present astounding and unique possibilities.

Yet ethological studies relating to *Orcinus orca* have only recently begun, and the bulk of these studies deal primarily with distribution, physiological, and predation analyses. Although orcas are found worldwide, little is known about the movements of any individual groups with respect to distances covered, extent of migrational movements, and so on. Intraspecific relationships, such as the means of communication or signaling, have scarcely been examined. Although there is some information on the structure of orca pods and the culmination of courtship behavior, our knowledge about the social organization of killer whale groups remains scanty. There is great need for further investigations.

In no other marine mammal do we have such convincing evidence of a complex communication system, together with an almost total ignorance of the content being communicated or the means of communication. As D. R. Martinez and E. Klinghammer (1970) state, "Signaling techniques and other means of communication, social interaction between members, the care and development of juveniles, courtship behavior and family organization, to the extent these phenomena exist, are fundamental aspects of behavior about which little is known."

From T. C. Poulter's (1968) study of killer whale vocalizations, he was led to conclude, "The system of signals of the killer whale seems to supply the strongest argument (from the standpoint of communication theory) that 'animal language' is statistically valid; we feel the killer whale signals are more important in this regard than are the signals of any other species. Yes! We believe that marine mammals talk, and that what they talk about makes sense to other marine mammals of the same species."

For all we know, the dolphins and whales may possess great minds, but at present we lack the technology to test and challenge those minds. This deficiency is matched by an absence of behavioral information.

We find the overwhelming barrier to the study of killer whale (and all cetacean) behavior to be an almost total lack of common experience between us, the land mammals, and the orcas, the sea mammals. Our meetings on their terms have usually been brief. They are gregarious creatures but highly protective of their privacy, and the sea is seldom cooperative. So we share little intimacy with them and their lives, and it is their life underwater that tells their story.

We can lower electronic instruments to track orcas and special optical devices to observe them at close range, but we feel that mechanical surveillance cannot equal our diving with killer whales in their own aquatic world. Few of us have had that opportunity, and fewer still have captured even a glimpse of wild, submerged orcas on film. This is a fundamental key in unlocking the secrets of this most unusual marine mammal.

Strategy

Our plan is simple. We intend to dive with wild killer whales and to obtain a film record of that interaction. This will take place in one of several locations in Washington and British Columbia especially selected for this project. (The location ultimately depends on the orcas.) That footage will be incorporated into a 45-minute documentary film on the world of killer whales in Washington and British Columbia.

We will charter a boat and equip it with all proper navigation, diving, filming, photographic, audio recording, and necessary living gear to approach killer whales and film them in safety. To attract killer whales, we will use three techniques to take advantage of their own natural curiosity.

1. A submersible, hand-held diver's sonar, which will be beamed at the orcas.

2. An underwater arc welder—the sparks fascinate orcas. (This was discovered by accident by a commercial diver friend of ours working on an underwater construction rig many years ago.)

3. A submersible, self-contained audio synthesizer, with stereo amplifier and public address system attached to the diver. This will be used to imitate vocalizations while in their presence underwater.

This last method of attracting and interacting with wild orcas has been quite successful. A synthesizer has been used (on the surface) to transmit orca vocalizations below the surface in an attempt to establish inter-species communication by using mutual imitation as a modality. It is clear that whales are highly interested in this experience; they will imitate synthesized signals and will present signals in what appears to be an attempt to induce humans to imitate them. This research needs to be done in a natural setting with complete recording of behavior as well as vocalization. Finally, an audio record will be made of all orcan vocalizations and diver reports given underwater as the events occur.

Only recently have we begun to understand killer whales as friendly, complex creatures who command our wonder and respect. A film document of the kind described in this proposal now seems timely and proper. We hope to accomplish the project during October 1977.

TRACKING ONE OF THE WORLD'S GREATEST UNDERGROUND RIVERS

Near-freezing cold. Total darkness. High humidity. Mud everywhere and in everything. And water. Water that can come rushing out of the inky blackness and kill by dislodging you from a perch on the edge of a hole 100 feet deep. Who would want to track this water to its most private lair, and why?

Any speleologist (or *spelunker* in American usage) could tell you the answers to that one. To go where humans have never been and to force a secret river to unfold geologic marvels that have never been seen is part of the answer. This project hints at some of the other reasons: persistence in tackling nature and persevering against conditions not designed to support human life. And there is more than a touch of challenging oneself, testing one's own nerve and capabilities.

JEAN-FRANÇOIS PERNETTE
Honorable Mention, Rolex Awards for Enterprise
"Pasquet"
33760 Escoussans
France
Born September 2, 1954. French.

A trilingual (French, Spanish, and English) student who earned his second-year license at the University of Bordeaux III, Jean-François Pernette is passionately involved in the world of speleology and is significantly contributing to its international expansion.

He has participated in expeditions in the Guadalupe Mountains of New Mexico and in the Black Hills of South Dakota (United States, 1974), in the Picos de Europa (Spain, 1973), in the Lamprechsofen (Salzburg, Austria, 1975), and in Belgium, Germany, Sicily, Switzerland, and, of course, France.

He participated in the first complete tour of the Pierre St. Martin cave, about which a movie was made. (And he was also a member of the four-man team who rescued the TV and police team lost for eight days while trying to make the same through-trip.)

Pernette is the French representative of the European section of the French National Speleological Society and is involved with many international speleological groups through correspondence and publications. Although he is young, the search for the St. Georges River has been an important part of his life.

I propose an expedition to further explore a huge cave system in the Pierre Saint Martin Plateau and to explore, survey, and study the recently discovered St. Georges underground river. To show the great interest of such exploration, I will offer some geographic and historical background, give a morphological description of the part of the cave that has already been explored, and detail some of the techniques we will be using.

Geographical Situation

The St. Georges River flows in the depths of the Anialarra Mountain, in the Pierre St. Martin Karst, in the provinces of Pyrénées-Atlantiques (French) and Navarra and Huesca (Spain). The entrance shaft, called

Pozo Estella, is in Spain, at a height of 2050 meters, while the resurgence (exit) is in Ste. Engrace (France), at 450 meters.

The Pierre St. Martin Karst is formed of a limestone mass that lies on impermeable schists. The whole plateau is tilted westward toward the resurgences. Because of the uniform geology, the profile of the caves is quite regular: The water permeates 400 meters of limestone until it reaches the impermeable schist and then follows the main faults, forming rivers of varying size.

The St. Georges River discharges 5.64 cubic meters per second at its resurgence. Its course has been charted by water tracing. Its depth potential is almost 1600 meters, and it has been explored down to -560 meters.

A few rivers similar to the St. Georges are already known. The most famous is the St. Vincent River, also called Gouffre de la Pierre St. Martin, which has an annual average discharge of 2.37 cubic meters per second and flows through the deepest cave in the world yet to be discovered.

Historical Background

In a hydrological sense, the St. Georges River is as famous as the Fontaine de Vaucluse. It remains one of the world's greatest hydrological enigmas. E. A. Martel and Max Cosyns were the first to notice its enormous potential, but like the other numerous expeditions organized from 1930 to 1970, theirs was unable to enter the hypothetical river. Meanwhile, several other rivers had been discovered at lower levels in the plateau.

Since 1971, I have been the leader of five expeditions searching for the river, each two months long, in which we descended over 5000 meters of virgin vertical shafts, some of them extremely deep, before we encountered the right one.

Description of the Cave

The entrance is a narrow, vertical chimney 150 meters deep that links the surface with a wide system of active pits. Then two pits, 25 and 33 meters deep respectively and separated by a narrow meander, lead to the only good platform of the vertical section. Ten meters below, a small streamway (discharge: $\frac{4}{5}$ liters/second) appears and flows down a narrow shaft. Nearby is another shaft through which water once flowed but that is now perfectly dry. Our exploration of this huge fossil shaft revealed two fantastic pits; one 100 meters deep, followed by a very narrow passage that leads to a huge pit 156 meters deep and completely blocked at the bottom by enormous rocks at a depth of 450 meters below the surface.

For further exploration in the active part, we made a small dam to divert the water into the dry pits. Below the dam, the cave continues, but the shafts are very narrow, with few ledges. This part of the cave is very dangerous; a flood has already claimed one of our teammates. After 260 meters of vertical descent (460 meters from the surface), we are now only at 65 meters, in a horizontal plane, from the entrance.

At the bottom of the vertical part, we reach the schist level, in a huge room through which a streamway flows westward. At its lowest level, its discharge is already about 200 liters/second. Upstream, the water runs into a wide passageway about 10 meters high and 30 meters wide. The slope is quite regular. This upstream part has been explored for 1 kilometer without any difficulty; we did not explore many of its branching galleries because we lacked time.

Downstream, the river disappears under the blocks of the first room and reappears on the other side of the room some 200 meters further. Then the water flows through the "Little Canyon" before cascading down a shaft 10 meters deep.

The next 2 kilometers are very irregular. The exploration becomes more difficult; the river forms either rapids or lakes, sometimes in such big rooms that one loses the notion of upstream and downstream because the river is under the blocks! In some narrow passages, a strong descending wind—proof, if needed, of continuation—shows the way and sometimes blows out our carbide flames.

Because we lacked time, food, and light, we stopped our exploration on top of a fantastic mound of rocks in a huge new room at a depth of 560 meters below the surface.

Techniques of Exploration

The difficulties of such an expedition begin aboveground. Expeditions must camp near the entrance shaft, which is two hours from the nearest gravel road. Moreover, the bad weather that is almost constant in the area, as well as the complete absence of water, exacerbate conditions.

The 460 meters of pits will be rigged with ropes. So far, all of the explorations have been possible in short operations of 20–30 hours, but now we probably must camp underground. The conditions in the cave are very peculiar: temperatures of 2–3° C, humid air so saturated that after a while camera flashes do not work, and mud and water that get into everything. We will install a telephone line between the surface camp and the first camp about 1 kilometer from the shaft's base. Via successive camps, we will then explore, map, and study the whole system.

We are now testing specific supplies able to resist water, mud, and humidity, as well as a variety of robust scientific instruments (compasses, chronometers for phone appointments and accurate sleep cycle studies, thermometers, hygrometers, water-tracing devices, and so on) capable not only of functioning in such bad conditions but also of surviving the rugged trip in the packs.

The remarkable geologic unity of the Pierre Saint Martin Plateau allows us to hope for success on the next expedition. The river is one of the world's most fascinating hydrological mysteries. Speleologists are aware that the system could join its small neighbor as one of the deepest caves in the world. At the same level, St. Georges River is far bigger than the Pierre St. Martin. And all the specialists agree that the descending wind, the size of the galleries, the morphology of the cave, and the geologic characteristics of the deep karst prove that the river continues past the point so far reached. We can only hope that we will be strong enough to conquer it. In any event, we'll do our best.

Anodorhynchus hyacinthinus, *the hyacinthine macaw.*

CAPTIVE BREEDING — LIFELINE TO THE FUTURE FOR ENDANGERED SPECIES

In spite of the best and considerable efforts of conservationists, winning the races to save wildlife species from extinction is by no means guaranteed. This is particularly true for many exotic species of birds, whose own natural habitats are being brutally destroyed and circumscribed by human encroachment. Although efforts are being made to stop this erosion of the birds' environments, it may well be that their future depends ultimately on our ability to preserve the species in captivity. Even with this secondary kind of conservation, there has been no guarantee of success, particularly in recent years when more and more legislation forbidding, or limiting, international animal trade has added to the problems.

As potentially the last repositories of exotic bird species, the zoos and aviaries of the world face the long-term challenge of maintaining their bird populations at, or above, replacement levels. Reliable methods for ensuring captive breeding programs are critical to the survival of many species.

In this project, an enterprising research endocrinologist at the San Diego Zoo seeks to broaden the base of a methodology that could become a standard for captive breeding management programs in zoos around the world.

BILLY LEE LASLEY
Rolex Laureate, Rolex Awards for Enterprise
Research Department
San Diego Zoo
P.O. Box 551
San Diego, California 92112
United States of America
Born June 4, 1941. American.

After taking his B.A. in life sciences (1963) from Chico State College in Chico, California, Bill Lasley went on to further education at the University of Washington, Emory University, and Oklahoma State University before completing his Ph.D. in physiology at the University of California in Davis. He is now adjunct assistant professor at the University of California in San Diego and research endocrinologist at the San Diego Zoo.

The primary objective of my proposed study is to verify and firmly establish a practical technique for sex determination of monomorphic birds. Preliminary studies in my laboratory indicate that fecal steroid analysis is a promising approach to the assessment of gonadal function in most, if not all, avian species.

Many species of birds are threatened with extinction owing to human intervention (habitat destruction, poaching, and pesticides), and captive breeding programs provide one of the best alternatives to their complete eradication. Although seriously attempted, breeding colonies frequently fail because of the inability to identify potential breeding pairs, especially when the sexes are monomorphic. Diminishing wild populations and increased barriers for international animal trade may mean that birds now in captivity may represent the ultimate future for many species.

The proposed project would provide the methodology and baseline data to enhance captive breeding of exotic birds. To be specific, a novel technique of evaluating reproductive status through fecal hormone analysis would become available to all interested parties. The proposed method has the advantages of simplicity, economy, and absence of stress for the birds. Although preliminary investigations have validated the technique for a large number of avian species, specific areas must now be researched, to provide the basis for broad acceptance and application of this methodology.

The greatest potential benefit of sex determination of monomorphic birds would be to enhance breeding management of rare or endangered monomorphic species. Applied prematurely, however, unforeseen species' differences or misapplication of the technique could be counterproductive. For example, erroneous sex assignment in only a few birds would further limit the captive breeding of a diminishing population. In this regard, we need well-defined studies to verify the application of this method to individual species. To do this properly, fecal steroid analysis must be correlated to gonadal histology at the time of autopsy, which is an unrealistic endeavor on a local basis. To meet this end, we propose that gonads and fecal samples of endangered birds that die be accumulated for evaluation. In this way, verification and application of the method could be rapidly approached.

Because there are limited numbers of some species, and even fewer that will be presented for autopsy, it is critical that the call for samples be as broad as possible; we hope it will be international. To accumulate these critical data, the San Diego Zoo Research Department proposes to call for and evaluate such samples as they become available. Gonadal histology, determined by standard techniques, will be correlated to fecal sex steroid values to provide a basis for the gonadal evaluation, using fecal steroid analysis alone, of living birds.

"*Samburu Girl,*" *Samburu District, Kenya.*

WE NEED PORTRAITS, NOT JUST PHOTOS, OF VANISHING CULTURES

The suggestion that we human beings are an endangered species is the first step in the attempt to keep ourselves from becoming extinct. Although we may be successful in that effort, we are losing all hope of preserving the variety of the human species. The distinctive flavors of our diverse cultural heritages are being lost in a homogeneous stew. Those who are concerned about our loss of these exotic ingredients realize that our children will be the poorer for not knowing strange lands and peoples that have captured the imaginations of all societies and ages.

Therefore, the attempt is made to document these lands and peoples. Masters of film and tape travel to the far corners of the world to capture records of sights and sounds that will cease to exist before our grandchildren are born. We can be grateful for these documents. But will they provide a portrait of these vanishing cultures, the sense of magic photographs often do not capture? If only daguerreotypes of Sitting Bull were left to us, and not his portrait, would our memory of him live as it does today?

This entrant thinks it would not, and she backs up her conviction by traveling to the vanishing peoples and painting them, in order to leave us with portraits of humanity, not just records.

LUNDA HOYLE GILL
6931 River Oaks Drive
McLean, Virginia 22101
United States of America
Born September 14, 1928. American.

A 1946 summer course in anatomy, design, composition, and drawing at the Chouinard Art Institute in Los Angeles is the first formal training listed by artist Lunda H. Gill. In 1950, she earned her B.A. degree in fine arts from Pomona College, Claremont, California, where she studied with Millard Sheets and Milford Zornes. Gill spent the next summer painting portraits with Jerry Farnsworth and still lifes with Helen Sawyer, at the Jerry Farnsworth School in Cape Cod, Massachusetts. Two years at the Art Students' League in New York City, studying figure painting with Robert Brackman and anatomy and oil painting with Edwin Dickinson, were followed by two years of drawing with illustrator Karl Godwin. She also studied at the Academia de Belli Arti in Florence and at the Frederick Taubes School in Sautillo, Mexico.

An ethnographic artist, Gill travels around the world to paint the peoples of vanishing cultures, from life and in their natural surroundings, recording them "before their magnificent beauty is lost forever to encroaching civilization." Her work has taken her to Kenya, to the homelands of American and Mexican Indians, and Alaska. The Smithsonian Institution has held a three-month showing of her Kenyan paintings.

The entire world is swiftly converging into one cultural pattern. Within ten years, cultures that offer the last living proofs of our cultural diversity will disappear. I must paint these peoples from life in their native environments to preserve and record their cultures before this occurs. Paintings draw us closer together, because art communicates in a way other media of communication do not.

I believe I am unique in my desire to paint most of the world's vanishing cultures. For this project, I wish to paint the people of Mongolia, New Guinea, and Nepal. I will videotape their stories and histories, make sketches and then return home to complete the paintings and write material to accompany the paintings. These paintings could be exhibited and published as art and educational material. With help and immediate action, the people of these extremely isolated cultures can be recorded and preserved in paintings.

My initial work in this field focused on American and Mexican Indians. Then, in 1974, I went to Kenya to research the tribal peoples of that country. This work culminated in a three-month exhibition of 35 oil paintings in the Smithsonian Institution's National Museum of Natural History in Washington, D.C. I am also writing a book on the Kenyan project. The third step was a trip to Alaska to record the Eskimos, Aleuts, and Indians; these paintings are still in progress. The fourth step is to paint the tribal people of Mongolia, New Guinea, and Nepal. I have asked the Russian government for permission to work in Outer Mongolia, and my application papers have been sent to Mongolia for official approval.

There is a real need to preserve for posterity rapidly vanishing ways of life. Some of the most informative and educational records available to us today are paintings, which, of course, also function as things of beauty. Many artists in the past have tried to record the cultures of their subjects; for example, George Catlin painted American Indians, and Rembrandt recorded the people of Amsterdam. Portraits of living human beings can record their inner essence (or soul) as well as the physical likeness. I find it miraculous to reach the point while I am painting when suddenly a human face looks at me from the canvas!

In the study of humanity, my work adds an extended dimension to the work of the anthropologist and the scientist. In 1966, the French anthropologist Claude Lévi-Strauss wrote:

> There are about 40,000 natives left in Australia, as opposed to 250,000 at the beginning of the nineteenth century. Between 1900 and 1950, over 90 tribes have been wiped out in Brazil; there are now barely 30 tribes living in a state of relative isolation. . . . Suppose for a moment that . . . an unknown planet was nearing the earth and would remain for 20 or 30 years at close range, afterward to disappear forever. In order to avail ourselves of this unique opportunity, neither effort nor money would be spared to build telescopes and satellites especially designed for the purpose. Should not the same be done [before it becomes] impossible forever?

With this challenge in mind, I have embarked on a program to record as many of the still surviving cultures as possible. It is inevitable they will change as technology creeps into their lives. I am not asking that technology be stopped, but I am asking to be a part of the recording, preserving, and studying, now, while it is still possible, before it is too late.

Moreover, as our world grows smaller day by day, we must learn to live closer to each other. I believe my paintings can help us in this endeavor. People are innately afraid of the unknown. When we see someone who looks, acts, or dresses differently from ourselves, we immediately draw back and put up our defenses. As we get to know about people different from ourselves, however, we usually begin to understand and like them.

Painting as a technique also overcomes another obstacle to recording vanishing cultures. Many people to this day are afraid of cameras and will not reveal themselves to the magical mechanical box. Because of the way painters work, they can break down this barrier and capture personality often not shown to the photographer.

Method of Work

I must handle my work in a very special way, because I am dealing with people so different from myself. I cannot bluster in and force people to pose, but must first gain their confidence, because there is absolutely no way to paint someone who is openly hostile. With a language guide, I try to move into a village quietly. Of course, this is not easy, as I look so different from the tribal people. Usually I just sit down, let my guide explain the situation, and let the people look me over. I also look them over, to choose subjects to pose for me. Selecting good subjects is of crucial importance and depends on my years of study, training, and painting hundreds of portraits.

When the people are finally at ease with me, after an hour or a day, I signal my guide to begin negotiations with the chief for the services of particular subjects. I try to choose faces that are representative of the tribe, and I make sure I get a mix of subjects—old, young, male, female, and so on. In my relationships with these people, I find it absolutely vital to be sensitive to their inner feelings and outer actions. It would be impossible to work if these people were not in total accord with me. I cannot paint what they do not give to me, so their friendship is a rare gift that I treasure.

I paint as many portraits as time and proper subject matter permit and then move on. After painting the field paintings, writing notes, and so on, I return home and paint final, detailed portraits. My exhibitions include both field and final paintings.

AN INTERNATIONAL COLOR STANDARD FOR BIOLOGY

How do you define a color to someone who is not looking at it? And why is it important that they know precisely what you mean? It is a maddening challenge, in spite of the availability of myriad printers' color charts.

For the biologist, who must describe a plant or an animal in precise terms, the problem of color is especially difficult. The growing interest of our era in protecting endangered species has added urgency to the search for better ways of identifying little known, and perhaps widely spread, flora and fauna. More accurate communication between the biological scientists around the world is clearly needed. The following project suggests a major contribution to that search.

J. HOWARD FRANK
Florida Medical Entomological Laboratory
P.O. Box 520
Vero Beach, Florida 32960
United States of America
Born April 13, 1942. British.

After taking his B.Sc. (honors) from the University of Durham in 1963, J. Howard Frank earned his Ph.D. at Oxford University in 1967. The two degrees, in zoology and entomology respectively, are part of the background he brings to his work as an entomologist in the Florida Department of Health and Rehabilitative Services. Additionally, his professional interests have led him into contact with the scientific societies connected with his project.

We cannot precisely indicate plant and animal colors, because we lack a good, modern, scientifically designed, low-cost color standard. The production of such a standard will assist in the description and identification of plants and animals and, by making them more easily identified by both scientist and layperson, will help us recognize and preserve rare and endangered species.

Species of mammals, birds, reptiles, certain plants, and certain insects (for example, some butterflies) in danger of extinction are frequently brought to the attention of the public in documentary films and magazines. The danger to these species is generally caused by human alteration of the environment. Environmental conservationists are making a praiseworthy effort, particularly in the developed countries and with regard to vertebrate animals and large plants.

However, vertebrate animals and large plants form only a small percentage of the total number of species of living organisms. The vast majority of species is only indirectly aided by the conservationists, that is, when an area is preserved in a natural or seminatural state. These areas may be selected because of their unusual scenic beauty or because they are the natural habitat of an endangered vertebrate animal or large plant. Yet the larger animals and plants are not logically of greater intrinsic worth than the smaller ones. The extinction of a species of beetle is just as great a loss as the extinction of a species of bird.

Why is not more effort made to conserve rare species of invertebrate animals and of the smaller plants? There are several reasons.

1. *Size.* The smaller animals and plants are generally less noticeable, so their existence is more readily overlooked by humans.

2. *Distribution.* Even in the more developed countries of the world, the distribution of these organisms has seldom been assessed thoroughly, so it is not well known which ones are in danger of extinction. In some of the developed countries, particularly in western Europe, intensive efforts are now being made to correct this lack.

3. *Identification.* There are so many species of invertebrate animals and smaller plants, and so few people (amateur or professional) able to identify them, that the general state of knowledge is far below that for vertebrate animals and cultivated plants. This problem is especially acute in the developing countries, where large numbers of species remain unknown to science and yet where wholesale destruction of habitats is occurring.

4. *Education.* It is evident that efforts to save the organisms from extinction, particularly in the developing countries, will never receive the necessary stimulus until the public can be educated to perceive the uniqueness of the organisms. This, in turn, cannot come about until the public can be provided with a name, a means of identification, and some biological facts about each organism concerned.

5. *Taxonomy.* The major problem underlying points 2, 3, and 4 lies in the relatively small number of scientists and amateurs describing and classifying these organisms. The public cannot be expected to take an interest in the organisms until it is provided with educational films and with illustrated guides to identification and to biology. These materials are costly to produce. Governmental agencies, more than any other sponsors, provide financial support for production of these materials, but in recent years the value of the support has declined, while the world's burgeoning human population has ensured the yet more rapid destruction of the environment.

Funding for a program to produce educational materials for the public, in order to protect the huge number of poorly known and undescribed animals and plants from extinction, would be enormously expensive. Funding for a program to protect one or a few species, or one or several geographic areas, would be feasible within the scope of the Rolex Awards but would be clearly inequitable in view of all the threatened species and all the ravaged habitats.

One particular area in which the Rolex Awards could help botanists and zoologists—amateurs and professionals of all countries—is in furthering the identification of all organisms, the common as well as the rare and endangered. This plan will protect the environment, if we consider plants and animals an integral and essential part of the environment, by advancing knowledge and understanding of flora and fauna.

In preparing a description of a plant or animal, a biologist uses words to describe its structure, color, similarities to other organisms, and size. In selecting the appropriate words, he or she draws on knowledge derived from many years of education, from textbooks, and from dictionaries. Moreover, biologists are trained to measure and express the measurements in figures. However, when it comes to describing colors, a biologist will soon be at a loss. The basic colors of the spectrum present no problem in nomenclature, but when a precise definition of a color from a range of even a few hundred is required, because the human eye can distinguish among several million different colors, it may be seen that the biologist needs a color dictionary. The reader of the description also needs a color dictionary to understand what is written.

There is, amazingly, no modern color dictionary in general use by biologists. Some biologists use no color standard at all in writing and reading descriptions of colors, while others use a variety of inadequate, poorly produced, faded standards with conflicting nomenclature. In brief, the situation is deplorable and has impeded and continues to impede the whole process of identification and accumulation of knowledge of organisms.

Activities and Plan

In 1973, the Coleopterists' Society (an international society for the study of beetles) established a color standards committee to examine the existing color dictionaries (standards) and advise the society about them. In 1975, this committee reported also to the Entomological Society of America (a much larger society for the study of all insects) on the apparent lack of a suitable color dictionary. A new committee, the Entomological Society of America Special Committee on Color Standards was formed, with five committee members interested in many groups of insects. This new committee found that other biologists were also interested in the availability of a color dictionary and that the Mycological Society of America (a society for the study of fungi) also had a Color Standards Committee, of six members. Members of these two committees found that their aims were identical and agreed to collaborate fully as a joint committee.

Members of the joint committees have examined copies of many color dictionaries, those currently available as well as those out of print and rare. These include, in chronological order of publication, Ridgway (1886, 1912), Saccardo (1894), Societé Francaise des Chrysantemistes et R. Oberthur (1905), Munsell Colour Company (1929, 1950), Maerz and Paul (1930), British Colour Council–Royal Horticultural Society (1938), Dade (1943), Villalobos-Dominguez and Villalobos (1947), Bielsaki (1957), Kelly and Judd (1965), Kornerup and Wanscher (1967), Rayner (1970), Stanley Gibbons (1973), Locquin (1975), and Smithe (1975). For various reasons, the joint committees consider none of these to be wholly suitable for use in taxonomic descriptions of plants or animals. If any one of them were useful, members of the committee would use it and save themselves further effort.

Collaboration between the Color Standards Committee of the Mycological Society of America (principally Dr. K. H. McKnight) and the US National Bureau of Standards (principally K. L. Kelly) has resulted in the development of a new color standard based on the ISCC–NBS centroid system for color designation, in which the color boundaries are defined precisely and colors are equidistant. A prototype of this new color standard is expected to be ready during the summer of 1977.

The joint committees are investigating a format and means of production of the new color standard, taking into consideration the following criteria: low cost, high quality, continuous availability, broad use, permanence, reproducibility, practical size, suitability for use in the field and laboratory, and complete color spectrum. They plan to use a simple numerical system to label the colors and to tabulate the names of the colors in English, French, German, Russian, Spanish, and perhaps other languages. Distribution and sales of the standard should be handled preferably by a nonprofit scientific organization, so it can be sold for the cost of production plus a minimal handling charge.

To date, the joint committees have not found a sponsor for the production of the new standard. A large sum of money will be required to enable the production of the first edition of 2000 copies. The parent societies do not have sufficient capital to act as sponsors. It is suggested that the initial capital would be returned to the sponsor (with interest) as sales progress. The production method would be formulated precisely so that second and subsequent printings would not differ from the first and so copies of the standard would be available for sale continuously in the foreseeable future.

Before. *After.*

RETRIEVING A WORLD FROM THE STONE AGE

The direction of civilization has not always been forward—witness the wreckage of a once lush countryside by human invasion of the Greater Chaco region of Argentina. Akin to the infamous Dust Bowl created in the United States, this once fertile land has receded into a primitive state barely capable of supporting inhabitants, both human and animal, at any level that could be considered much further advanced than that of our prehistoric ancestors.

The scrub desolation of the Greater Chaco in Argentina is not, however, destined to continue forever. Viewed from above, its rolling miles of wasteland would seem to be incapable of change—until one spots a patch of verdant green like the Garden of Eden. What happened? The answer to that follows, and in the answer is an object lesson for us; one highly motivated individual can change the face of the earth for the better.

JORGE SAMUEL MOLINA BUCK
Honorable Mention, Rolex Awards for Enterprise
Research Center on Microbial Ecology
C. P. 1428, Obligado 2490
Buenos Aires
Republic of Argentina
Born June 29, 1919. Argentine.

As full professor on the faculty of agronomy at Buenos Aires University, Jorge S. M. Buck lists a long line of public and professional credits and awards for his distinguished work in Argentina. He is also the head of the Research Center on Microbial Ecology, whose principal investigations are related to erosion control, nitrogen fixation, and the reclamation of alkaline soils by biological methods.

An agricultural engineer by education (Buenos Aires University, 1943), Professor Buck was awarded an honors diploma there, the first of several awards he has earned, many of which are national in scope. He has been the driving force behind a project on a startlingly large scale—both for national implications and those beyond the borders of Argentina.

I hope to transform 20 million hectares (200,000 square kilometers) of presently unproductive land in the semiarid region of the Argentine part of the Greater Chaco. The principal goal is to produce bovine meat (animal protein) on a scale of 200,000 tons (minimum) annually and to grow numerous subproducts, such as sorghum and seed of typical forages.

The feasibility of this project has been demonstrated in a pilot area of 53,000 hectares, and we estimate our success can have a considerable impact on worldwide animal protein production. The importance of our study rests not only on the meat produced, but also on the development of new techniques that could be applied immediately to other regions of the world with similar problems. In the Paraguayan and Bolivian parts of the Greater Chaco, which cover about 40-50 million additional hectares, similar projects are proceeding under the supervision of former assistants at Buenos Aires University.

Background

The military conquest of the Greater Chaco (1878-1880) took place almost simultaneously with the conquest of the Pampas. In places called

Resistencia, Reconquista, and so on, whose names indicate the difficulties of the operation, a line of forts was built that led to the establishment of many of the present towns and cities in the area. After an initial stage of forest exploitation (tannin extraction), those developing the agricultural possibilities encountered enormous obstacles, which finally discouraged even the most daring.

In the first place, locusts regularly devastated all the crops. Also, the kinds of crops were poorly selected; wheat was preferred, despite the problems of being in a marginal area, great distances to the shipping port, the lack of adequate varieties, and so on. There was, and still exists, an almost complete ignorance of the dry farming techniques so essential to a semiarid zone.

In these conditions, complete failure was normal, and the few survivors were forced to dedicate themselves to animal husbandry. But in these conditions even animal husbandry had certain limitations. The various types of English stock, practically the only ones available, were neither adequate nor suited to the subtropical conditions of the semiarid Chaco. The more crossed the strains became, the lower production fell. Furthermore, the limited supply of water for the animals was in itself a practically insurmountable problem. Even the windmill techniques so useful in the Pampas were not adequate for the Greater Chaco.

Given these inconveniences, coupled with the abundant dispersal of agricultural land in the Pampas and the sparse population of the Argentine Republic, there was little incentive for a concentrated effort to solve the problems of the Greater Chaco. Under those conditions, in less than a generation, fields were abandoned, and scrub forest covered the area once again.

Today, conditions have changed radically, and new plans appear promising. A few of the changes are noteworthy here.

The Locust. As a consequence of the integrated campaign against it, involving massive use of new insecticides, the locust is no longer a problem.

The Utilization of the Land. Any purely cash crop system will, in this region, be a failure. The risk implied in cash crops is too big to be the basis of exploitation. Instead, to complement a well-oriented grassland farming system, selected crops are quite useful.

The principal cash crop associated with animal husbandry is sorghum. Its rational use in rotation with a legume, such as sweet clover (*Melilotus alba*), can ensure excellent yields in years of good rains. In dry years, it is used as a pasture for animals. In this zone of very little frost, sorghum can provide nearly nine months of grazing for cattle.

Sweet clover is really an extraordinary legume for the zone. It is so well adapted that it can be found within the forest scrub, competing with native plants. Its capacity for producing food of high protein content in full winter makes it totally irreplaceable as a basic element of the grazing chain; it complements sorghum at precisely the time that sorghum stops producing.

Another good pasture is Rhodes grass (*Chloris gayana*), which could provide permanent pasture for the breeding animals. Weeping love grass (*Eragrostis curvula*) can be an excellent crop for the sandy soils present in many regions. Other, less studied possibilities that could be useful in the rotation schedule are cowpeas, soybeans, and sunflowers. In all cases, the clue to every successful crop in this zone is a good previous fallow, which notably reduces the risks of scanty rainfall.

Cattle Breeds. A fundamental change in animal husbandry planning consists of the replacement of all animals of British origin with Cebu animals of Indian origin. The most promising variety is Nelore, which in Brazil has replaced all other Cebu types.

Water. Lastly, water is the fundamental ingredient for success or failure in all agricultural projects in the Greater Chaco. The traditional irrigation systems of the Pampas failed in the Greater Chaco. Generally there are no deposits of drinkable water at the subsoil level, or they are of limited volume. This was the nemesis of all animal husbandry in the zone.

Interesting data about the situation could be obtained in the Paraguayan Chaco, so we organized a tour with Paraguayan ranchers and learned that Paraguay, with the aid of some notable international institutions, had solved the water problem in the northern Chaco. Ranches of the Paraguayan Chaco, even those of 40,000 hectares and 22,000 cattle, and with eight months of drought, had no water shortage for cattle. The secret consisted of a network of dams that collected water during the rainy season (summer) and stored it for the drought season (winter). With an average volume of 10,000 cubic meters and a depth of 4 meters, an adequate conservation of water was achieved without the use of any subsoil water.

This technique on an ample scale in the southern Chaco (Santiago del Estero Province) gave the same results. We in Argentina have an additional advantage, because we can periodically use subsoil water. A proper mixture of saline water from the subsoil and nearly distilled water from dams offers an optimal quality and quantity of water for different types of cattle. A special plastic densimeter was invented to calibrate in the field the correct saline concentration of the water mixture.

Detailed Description of the Project

In May 1965, we published an article in the magazine *Dinamica Rural* entitled, "The Country of Wilderness: A Myth to Be Destroyed," in which we indicated our convictions that the semiarid Greater Chaco held great agricultural possibilities. Later, in the Buenos Aires newspaper *Clarin*, we published the plan for an "Agricultural Conquest of the Greater Chaco."

The interest that these articles aroused persuaded us to publish a series of articles in which we summarized our experience of nearly 30 years in central and southern Chaco and detailed the geopolitical importance of this enterprise. At the request of the government of the Province of Santiago del Estero, we undertook the exploration and study of the agricultural possibilities of the southern Chaco in an area of approximately 2 million hectares. In 1970, a private company called Estancias S. A. asked us to supervise the development of 53,000 hectares. This area comprises four different *estancias* (ranches) that encompass the variety of physical situations in the Greater Chaco.

This pilot experience permitted us to solve the last problems that handicapped development of the region. Drinkable water for people and cattle was provided in a region where no drinkable water was in the subsoil. We also accomplished the following: We accelerated elimination of all unproductive bushes and forest at the rate of 4000–5000 hectares per year; reclaimed alkaline soils and transformed them into grazing land with high production of bovine meat; resolved the problem of races and crosses of cattle adapted specially to the region; adapted the forages to the extreme conditions of the zone; established methods of soil management that permit high production and also preserve soil fertility; employed systems of rotation that avoid soil exhaustion and erosion; developed agricultural machinery that permits work in recently cleared soils; and employed mechanical methods that avoid the need for seasonal burn of natural grasses, which is a true scourge of all semiarid regions. At the same time, our scientific research developed the theoretical basis of all these practical applications and their reliable application to other similar semiarid tropical and subtropical regions.

The practical results of all this work can be summarized and explained with the following data, taken from the pilot operation:

Before	*After*
1 cow each 8–10 hectares, in bad condition	1 cow each 1.5 hectares, in excellent condition
Poor grazing land in summer and no grazing land in winter	In summer, 2395 hectares of Rhodes grass, 2348 of sorghum; in winter, 2218 of sweet clover
Only one good watering station	40 watering stations (dams) of 8–10,000 m^3 each
Low meat production—only a few kilos/hectare/year	Good meat production—more than 100–150 kilos/hectare/year
Zero agricultural production	2000 hectares sorghum, 30–40 quintals/hectare 1000 hectares sweet clover for seed production, 200 kilos/hectare
Only 30–40 percent parturition	More than 70 percent parturition
No steer production, only calves	Steer production 400 kilos in a little over two years
Only a few kilometers of wire fence	More than 100 kilometers of new wire fencing

Basic Research

Recovery of Alkaline Soils. Thanks to basic research in laboratories, we could, using biological methods, transform sterile alkaline land into pastures of high meat productivity. The practical application of this research permitted the recovery of more than 350,000 hectares of formerly unproductive soils.

Improvement of Soil Structure and Struggle Against Erosion. The aerobic decomposition of cellulose during the production of polyuronic colloids (which act as soil conditioners) permitted us to establish a new theory of soil conservation. In practice, the theory led to improved soil structure and avoided erosion of nearly 2 million hectares.

Nitrogen Fixation by Nonsymbiotic Bacteria. The theoretical studies are nearly complete. We still need to find a practical method for applying these studies in the field. The theoretical possibilities can be deduced from the fixation data obtained through gas chromatography (55.9 kilos/hectare/day). These research data were gathered in Argentina by Santos Soriano and Amor Asuncion.

Measurement of Water Salinity. A plastic densimeter was developed that permits measurement of a mixture of nearly distilled water from dams with the saline water from wells, thereby achieving an economy of potable water for both humans and cattle.

Despite the problems encountered in the Greater Chaco, it has been possible, by applying the most modern agricultural technology, to transform 53,000 hectares in only three years (since 1969) and to obtain productivity levels similar to those of the Pampas region.

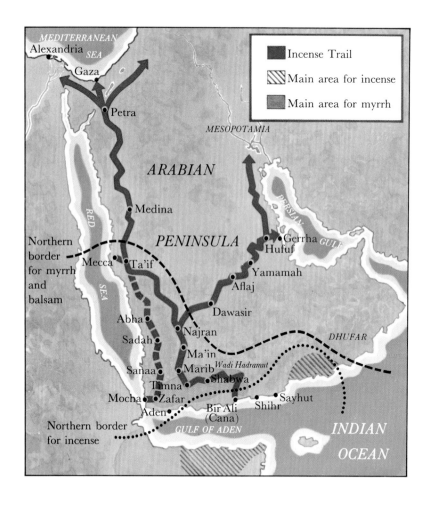

BEFORE OIL, THE TRAIL
LED TO INCENSE

Well over a thousand years before the great pipelines began snaking their dull, metallic tentacles across the inhospitable sands of the Arabian Peninsula to feed the insatiable demands of alien lands for ever more oil, an earlier, livelier, and more exotic set of routes brought prosperity to the peoples of the Arab world.

Endless caravans criss-crossed and skirted the deserts, in response to an extraordinarily long-lived demand for incense and myrrh. Seaports, cities, and cultures grew along the paths of camels laden with antiquity's commercial equivalent of today's oil.

Perhaps the most famous of these routes, nearly lost to us today, is described here. If its traces were forever to disappear, we would lose much of our ability to learn and understand the background of the peoples of Arabia. The ancient place-names remain, as do the unraveled tensions and conflicts, for the modern explorer who cares to seek them. And perhaps the search, as proposed here, *should* be done from camelback!

HUGO FRANZ BAUR
Honorable Mention, Rolex Awards for Enterprise
Balsbergweg 20
CH-8302 Kloten
Switzerland
Born December 7, 1941. Swiss.

Hugo Baur, an administrative assistant with Swissair in Zurich, has traveled widely in the Middle East, Asia, and Africa—mostly off the beaten track. His knowledge of English, German, and French provided background for his study of modern literary Arabic at Zurich University. He feels his course studies there and his familiarity with the customs and life patterns in Arabia will be of particular value in the exotic expedition he proposes here.

The objective of this one-man expedition will be to retrace the route of the ancient Incense Trail in Arabia from Bir Ali (ancient Cana), the most important ancient seaport on the Arabian Sea (in today's People's Democratic Republic of Yemen), to Petra, in Jordan.

The route will follow the ancient Incense Trail, which can be ascertained with reasonable accuracy from various sources. The expedition will start in Bir Ali and will proceed to the Wadi Hadramut (Tarim, Shibam) and on to Shabwa, the actual starting point of the Incense Trail. On its way to Petra, Jordan, the expedition will take the following route: Shabwa to Timna (ancient capital of the Qatabanians) to Harib to Marib (capital of the Sabaeans) to Ma'in (an important city of the ancient Minaeans) to Najran to Ta'if to Medina (old Yathrib) and finally to Petra.

This will be a one-man expedition, lasting 90 to 100 days. Local guides, however, will be hired for the largest part of the expedition. The first leg, from Bir Ali to the Wadi Hadramut, will be by hired motor vehicle. From the Wadi Hadramut to Petra, camels will be hired locally for periods of 5 to 10 days, together with a guide and one or more helpers.

During the entire expedition, special attention will be given to assembling a written record (together with supporting photographic material) of the modern environment and features of the ancient Incense Trail. Detailed notes will be made in the following disciplines:

geography and geology (the geographic and geologic setting of the Incense Trail), social anthropology and ethnology (notes on the tribes who now live along the Trail), linguistics (studies of the modern dialects of Arabic spoken along the Incense Trail), miscellaneous (collection of data relating to the climatic conditions prevailing during the expedition, and rough mapping of poorly or insufficiently mapped areas).

The importance of the ancient Incense Trail is emphasized in the *Westerman Lexikon der Geographie* (see entry "Incense Trail"), translation by Hugo Baur:

[The] Incense Trail [is] one of the oldest routes of world trade. From the third century B.C., right through the seventh or eighth centuries A.D., the Incense Trail connected several seaports of Southwestern Arabia between Dhufar and Muza (near the present-day Mocha, in Yemen) with Gaza on the Mediterranean. Its importance emanated from the trade in incense and myrrh which were grown on the shores of the Gulf of Aden. The Incense Trail furthermore acted as an important vehicle for the exchange of the cultures and civilizations of the Mediterranean on the one hand, and of India, South East Asia, China and East Africa on the other. According to Pliny (died 79 A.D.), the Roman Empire imported goods worth 100 million sestertia every year from Arabia, India and China.

This old caravan route had important feeder trails from Dhufar, Sayhut, Shihr, Cana and Eudaimon Arabia (Aden). Its starting point was at Shabwa in the Wadi Hadramut, and it led through the Wadi Maifaa to the famous royal cities of the eastern foothills of the Yemeni highlands: Timna (Qataban), Marib (Saba), Ma'in (Minaeans). At Najran, one route branched off via Dawasir, Aflaj, Yamamah, Hufuf, to the Gulf and Mesopotamia, whereas the main trail led to Petra via Turabeh and Yathrib (modern Medina). From Petra, various routes continued to Gaza on the Mediterranean, and to Egypt and Syria. The main entrepôt port in the Mediterranean was Alexandria; for the trade with India, Mesopotamia and the Gulf area, it was Gerrha.

Wars over the control of the Incense Trail resulted in the decline of this very lucrative route of trade. Soon a second route through the Yemeni highlands was in use, parallel to the original Incense Trail. This route connected Zafar, Sanaa, Sadah, Abha, Ta'if with Mecca. It was partly paved, and is still used today by caravans.

The decline of the kingdoms of Southern Arabia, and the rise of maritime transportation in the Red Sea, led to the abandonment of the Incense Trail. The incense from Africa was shipped around Southern Arabia, and was delivered to Jiddah, and the Persians began to dominate the entrepôt trade with South East Asia. This resulted in deterioration in the state of the economy of Southwestern Arabia. The extensive systems of irrigation fell into disuse, and the famous dam of Marib was destroyed towards the end of the sixth century.

AN 80-FOOT STEEL CATAMARAN: ONE-MAN CRAFT FOR ONE-MAN CREW

Few shipyards, let alone individuals, would take on the job of building an 80-foot steel catamaran, much less attempt to design one that could be sailed by one man. The mathematics of sail handling at sea would seem to imply that such a craft was beyond all reasonable aspirations.

Not so, says a mathematics and science teacher from Australia, a land where sailing is taken very seriously and where one does not venture nautical opinions unless ready to back them up. Backing up this opinion is what this project is all about, and the boat that painstakingly and laboriously has been coming together over the past four years is direct testimony to the will of one individual. She will be regal, and she will be royally free to roam the seas of the world at will, owing fealty to no one but the man who conceived her and built her and who can sail her by himself.

LESLIE GEORGE THOMPSON
83 Halford Street
Inverloch
Victoria 3996
Australia
Born October 17, 1935. Australian.

Leslie G. Thompson completed his teacher training at the University of Melbourne and State College of Victoria during the years 1963–67, prior to becoming a secondary school teacher of science and mathematics. With an apprenticeship in fitting and turning completed in 1955 and with ten years of industrial experience as a marine engineer, fitter, and draftsman, Thompson, not surprisingly, is in a position to take on the following enterprise, which few people could handle.

The steel craft is designed to be comfortable, useful, fast, safe, and ocean-going with shallow draft. It is 24 meters (80 feet) long, 10.3 meters (33.8 feet) wide, and 6 meters (19.7 feet) high (excluding masts). She is designed to be sailed by one man.

I have been attracted to catamarans since I was a youth. Various projects have been undertaken and much learned before I undertook this one. I decided to build the craft 24 meters (80 feet) long to allow headroom in the wing. Thus the accommodation is increased immensely, as is the safety, since the overturning moment increases as the third power of the length, while the restoring moment increases as the fourth power.

Catamarans are inherently fast, being free from the fundamental speed limit imposed on a monohull owing to its beam. The hulls on my craft are slim (aspect ratio 10:1) and are parabolic in section (which means a minimum wetted surface area). The design of the catamaran, which is entirely my own, was first conceived in 1967. First, I designed it and built a model. Then I built the necessary machinery: forklift, bench shears, drilling machine, grinder, sandblasting machine, and power hacksaw. In February 1972, I began construction of the actual craft; now the steel work is nearly complete. All this has been done as a spare-time activity in conjunction with my job, family, and home commitments.

Much of the design is without precedent; large steel catamarans are uncommon. The hulls are highly compartmentalized. If water creeps into any portion of a hull, it cannot progress along the hull for more than one-quarter of its length, because the bulkheads have no openings in the hulls. Furthermore, the lower decks, which constitute a double bottom, are in each compartment except the engine room. For its size, the craft is extremely light and has a shallow draft, compared with a monohull.

The design of mast, sails and sail handling, and centerboard is uncertain as yet, but I plan that these will allow the vessel to be managed entirely by one person. I also hope to make the mast retractable, or lowerable, while at sea, and in this regard, as with the sails, the great beam of a catamaran is of immense value. The design concepts may be summarized as follows: semirotatable wing mast, using sail track and conventional sails; fixed-angle, retractable mast, using jib-type sails for all of them (no booms or sail tracks); mast of steel tubular space frame, in two sections, to telescope (hinge down) and lie flat on deck; centerboard surface pierced with fence to inhibit air entrainment, mounted under wing between hulls; and centerboard hinged to allow shoalwater sailing, increased safety in high seas, and adjustment of center of lateral resistance, for optimum sail balance.

Construction

Welded steel. Plated with $\frac{3}{16}$ inch mild steel below the water line, and $\frac{1}{8}$ inch above. Framing is $2\frac{1}{2}$ inch $\times \frac{1}{4}$ inch, or $2\frac{1}{2}$ inch $\times \frac{1}{8}$ inch in the hulls, and 2 inch $\times \frac{3}{16}$ inch in the wing. The bulkheads in the wing are integrated with those in the hulls. This feature provides great strength and allows use of the wing for accommodation. Where stress concentrations arise because of reentrant corners (for example, the bottom of the wing and inner side of hulls), gussets are provided.

The entire exterior surface is being sandblasted to white metal and painted with zinc silicate primer, followed by further coats of suitable types of paint.

Length: 24 meters (80 feet)   *Beam*: 10.3 meters (33.8 feet)

Draft: 1.2 meters (3.9 feet)   *Displacement*: 57 metric tons (62.8 tons)

Sail Area: 200 square meters (2160 square feet)

Two Auxiliary Engines: diesel, each approximately 75 kilowatts

Sail Description: Ketch rig

Two Masts: Steel space frame construction, using elliptical section tubing

LOOKING FOR THE KEYS TO THE DISARMAMENT/ARMS CONTROL PROBLEM

When this project's author speaks of humans as an "endangered species," he does so with the background of long involvement with one of the nastiest problems we face today: how to prevent blowing ourselves off the face of this planet with the aid of our own brilliant nuclear technology.

In our long history of attempts to limit war, or at least the worst ravages of armed conflict, humanity has never had to face such "ultimate" weapons as those with which we grapple today. Even prior to the time when a long-distance weapon could destroy a faraway society, we never achieved lasting progress in limiting our ability or willingness to kill each other.

There have not been any good solutions to arms control and disarmament, but then the stakes on the table have never been quite what they are today. The escalating arms races, nuclear proliferation, and rapidly shifting tension points—all warrant gloom. The kind of negotiations on arms that took place in the past (and were found lacking, even in those less cataclysmic times) desperately need to be replaced with a better alternative. The search for that alternative is the subject of this project.

WILLIAM EPSTEIN
Honorable Mention, Rolex Awards for Enterprise
United Nations Institute for Training and Research
801 UN Plaza
New York, New York 10017
United States of America
Born July 10, 1912. Canadian.

After receiving his bachelor of arts (1933) and his bachelor of law (1935), both from the University of Alberta, William Epstein went on to postgraduate work at the London School of Economics during 1937–38. Following World War II military service, in which he rose to the rank of captain, he became an official in the Secretariat of the United Nations in 1946. Since retiring in 1973 from his long-time position as director of the Disarmament Division of the UN Secretariat, Epstein has come to hold two posts, both in the field in which he has worked for over 25 years. He is an honorary special fellow at the UN Institute for Training and Research in New York, where he is responsible for research in arms control and disarmament. He also was appointed visiting professor at the University of Victoria, Canada, where he gives lectures and seminars on international security, arms control, and disarmament. He has written extensively, in books and articles on the subject with which his project is concerned.

The ultimate and greatest danger to the environment, the threat of a nuclear holocaust, is growing, not diminishing. Because all past efforts at arms control and disarmament have failed, we must seek new approaches.

I propose to undertake a study of new substantive and institutional approaches to arms control and disarmament. The project would explore why and how previous efforts have failed, whether a new, comprehensive approach to both nuclear and conventional armaments might be more successful than the step-by-step or incremental approach, and whether the reactivation of the United Nations as the main forum for disarmament negotiations, and the consequent involvement of China, France, and developing Third World countries in the negotiations, might produce better results. I would propose models for both short- and long-term disarmament programs, as well as new institutional and procedural mechanisms to facilitate real disarmament.

Successful new approaches to disarmament are desperately needed if the world is to be saved from the horrors of nuclear war and if humanity is to survive. My project might tentatively be entitled, "Humanity Is an Endangered Species" and subtitled, "New Approaches to Disarmament and Human Survival."

The project would involve a detailed exploration and testing of possibilities for real progress in three areas.

First, a comprehensive, substantive approach would offer larger packages of both nuclear and conventional measures. The incremental approach has failed in large part because arms control negotiations move much more slowly than do weapons technology and production. It may be more difficult to arrive at a fair "balance" if each weapons system is dealt with separately, because each system is linked with all others, and whatever is done about one has an impact on the others. Clearly, little progress can be made unless we consider nuclear and conventional disarmament together. It may therefore be easier to achieve acceptable balance at lower levels if larger packages of nuclear and/or conventional disarmament measures were formulated.

A broader mix might also reduce interagency rivalries. In addition, comprehensive measures, as well as specific measures of control and verification, could conceivably reinforce the arms balance. A comprehensive program might provoke less fear of temporary imbalance and might generate greater momentum.

With a comprehensive program, it might be easier to initiate unilateral arms limitations and restraints, which might result in greater openness, exchange of information, and trust. The study would explore and assess the feasibility, nature, and extent of such initiatives. It would also explore whether a comprehensive program of disarmament would attract needed public interest and have a positive influence on the highest levels of government. The very fact that new, far-reaching approaches were being considered might stimulate greater interest, debate, and study, which in turn might lead to breakthroughs.

The study would examine and elaborate short- and long-term models for comprehensive disarmament programs.

The short-term models, covering a period of three years, would include "packages" of various measures for "freezing" the arms race: (1) halting underground nuclear testing by the United States and the USSR, (2) trial suspension by the United States and the USSR of flight-testing new or improved nuclear missiles and delivery vehicles, (3) 10 percent reduction of nuclear weapons for each of the three years by the United States and the USSR, (4) achievement of substantial parity in NATO and Warsaw Pact conventional forces, (5) pledges by the United States and the USSR (and other nuclear powers) not to use nuclear weapons against nonnuclear powers and not to initiate nuclear weapons against other nuclear powers, (6) suspending export of conventional arms pending negotiation of agreements for drastically reduced exports, (7) freezing military budgets at current levels, and (8) commencing negotiations for long-term disarmament.

The long-term models, covering a period of about 15 years, would include alternative measures for reducing nuclear armaments of the United States and the USSR to about 10 percent of present strength. This would result in reductions by other nuclear powers, as well as in drastic reductions in conventional armaments and military expenditures by all countries, to about 10 to 20 percent of present levels. Control and verification measures, mainly national and technical, would be analyzed to see how much control is necessary. A progressive strengthening of the United Nations' peace measures would also be outlined.

Second, the project would offer a more active role for the United Nations. The study would examine whether universal participation, involving more countries in the negotiations, might work better than other bilateral, regional, or functional approaches. Involving all interested countries might stimulate the two superpowers. At the same time, directly involving the Third World and developing countries might make it easier to persuade them of the benefits of restraint in acquiring nuclear weapons or large stocks of conventional arms. Nonproliferation policies might become more credible, and perhaps more acceptable, if the potential proliferators became seriously involved in disarmament efforts. Although not all 147 members of the United Nations can participate in all of the negotiations, they could participate in all discussions in a more meaningful way than in the present, limited General Assembly sessions. The real negotiations would, as always, proceed in small groups and in delegation offices of the countries concerned. The broader discussions might also educate both rich and poor countries concerning the interrelationships among the arms race, disarmament, and socioeconomic development. Moreover, China and France, which boycott the Geneva Conference of the Committee on Disarmament but did not oppose a special session of the UN General Assembly on Disarmament, would find it difficult to boycott UN disarmament efforts.

Third, new institutional and procedural mechanisms would be explored. Although none would be simple to implement, many new approaches can be visualized, including the following. (1) An annual report by the UN Secretary General on the arms race and on disarmament (similar to his annual world economic and world social reports). (2) UN publication of periodic reports (monthly or quarterly) of all developments in the armament and disarmament fields (it has been decided that the United Nations will publish an armaments yearbook). (3) UN creation of an arms review committee that would review the necessity and impact of developments in the field of armaments and military expenditures. This committee might function in some ways similar to the UN Decolonization Committee and the OECD's economic review bodies. While other new disarmament bodies (including France and China) discussed the problems of "disarming down," the arms review committee would discuss the problems of "arming up." (4) The establishment of a satellite surveillance system, by the United Nations or some subsidiary body, that would provide information to all countries, to increase the openness of information and international confidence. (5) The eventual establishment of a new UN agency for arms control and disarmament, along the lines of the existing specialized agencies.

Research Plan and Technique

In carrying out the research, I would build on my previous studies of arms control and disarmament, including work on nuclear proliferation and comprehensive disarmament. The proposed program is more ambitious than my previous work and would require a research assistant.

Library and documentary research at the United Nations and the Library of Congress would be supplemented from my own voluminous papers, memoranda, diaries, and notes. It would also involve personal interviews with government officials, US Congressmen and their staff assistants, UN delegations in New York and at the Geneva Conference of the Committee on Disarmament, with nongoverment scholars and analysts (mainly in the United States but also abroad), and with Soviet scholars at Pugwash and other international conferences. In these interviews I would not only gain information but would also test various hypotheses and my own ideas. Because some of these ideas are rather novel and far-reaching, I would circulate portions of my work to both officials and scholars to elicit their reactions, and I would also try to arrange to discuss these ideas at seminars both at universities studying arms control and elsewhere.

The Sudd Swamp, Upper Sudan.

MONITORING THE WORLD'S LARGEST BIOMASS — BEFORE THE CANAL COMES THROUGH

Our great building projects are not normally designed to damage our environment, and, indeed, they often have a negligible effect. Other constructions, such as the Alaskan pipeline, can wreak havoc on delicate balances between flora and fauna, on soil vitality, and the like.

In similar fashion, a huge new canal, carved through a territory holding the largest number of herbivores in the world, poses major upheavals in a natural ecosystem so vast it has never been more than cursorily examined and understood.

In a race against time, a highly skilled and competent ecologist is attempting to provide authorities with critically needed information about the wildlife and terrain that the canal will affect. It is likely that the animals involved have no greater champion in the world than this concerned individual.

ROBERT MURRAY WATSON
Resource Management & Research
P.O. Box 561
Khartoum
Sudan
Born February 17, 1939. British.

Robert M. Watson's early education was at the Lincoln School and Archbishop Holgate's School. He won a state scholarship and a North Riding County Major Scholarship, both to Cambridge University, in 1958. He has received several awards, including one from the Colonial Office of London in 1960 as the leader of the Cambridge expedition to British Guiana (now Guyana).

He took his B.A. (honors) in zoology in 1961 and his M.A. in 1966 from Cambridge University. He was awarded his Ph.D. in 1967 by Cambridge University for his thesis, "The Population Ecology of Wildebeeste in the Serengeti." In 1970, he received the Cuthbert Peake Award, given by the Royal Geographical Society for work with the Society's South Turkana expedition.

As founder and senior partner of a consulting firm working in Africa on problems involving natural resources and their development according to ecologically sound principles, he performs a wide-ranging role. A pilot with 10,000 hours bush flying experience, he uses a low-level aerial sampling technique, of which he was the principal developer, to provide information on natural resources rapidly, precisely, and cheaply.

I propose an ecological investigation of the Sudd Swamp and its floodplain to establish its chief physical and biotic parameters. The Sudd is the largest freshwater swamp in the world. Taken in conjunction with its surrounding floodplain, the area supports the largest wild herbivore population in the world. Biomasses in the Sudd and its floodplain have been estimated at between two and three times those of the Serengeti region, its nearest rival.

The government of Sudan has embarked on a program to construct the Jonglei Canal, which could alter the water balances, and hence the whole ecology, of the area. At present, there are no quantitative data on the physical and biological characteristics of the region that could be used either to assess the possible impact of the Jonglei Canal developments or to monitor such changes as do take place as a result of the canal's construction, in response to other developments, or as part of the natural dynamics of the ecosystem.

I propose to conduct an ecological study in the area that would serve three principal functions. First, identify and quantify the major components of the ecosystem that should be monitored over succeeding years in order to assess the trends and fluctuations of the system. Second, establish in quantifiable terms and in detail the present nature and state of this unique swamp and floodplain ecosystem. Third, test a methodology that will enable a continuous monitoring of the area's conditions, so that a reliable means of predicting changes will be possible.

Secondary or subsidiary benefits of such a study would provide: (1) "early warning" of impending large-scale ecological changes in the region; (2) sound data on which an optimization of alternative forms of land use may be made. Alternatives include use of the area or parts of it for recreational purposes (which would entail creation of reserves and national parks), use by subsistence crop producers and grazers, and use for larger crop and grazing schemes; (3) adequate data to ensure the survival of the unique components of the system, such as the Nile Lechwe, under whatever development program is eventually pursued.

The method of undertaking the work hinges on an interesting form of sampling, in which physical and biotic profiles will be quantified and recorded at statistically predetermined positions across the Sudd floodplain system. The location of these sample profiles will be fixed by using aerial photography at different scales, marker buoys, and beacons so it will always be possible to return to exactly the same sample. The physical and biological profiling of the samples will be carried out in Land Rovers, work boats, and light aircraft. One of these aircraft will be fitted with floats so we can reach and sample inaccessible lagoons deep in the Sudd that have never been visited by people.

Measurements and quantifications will be made with the normal range of hydrological and terrestrial survey instruments, low-level aerial photography, underwater photography and diving, the company's own low-level aerial census methods, and others. New techniques of sampling aquatic flora and fauna and the flow patterns in the Sudd will be developed to provide rapid, cheap repetition of measurements in future years. Through investigations of mud deposits, the distribution and nature of papyrus and reed beds, and a comparison of old aerial photographs with current photographs, we hope to determine both the recent and ancient ecological and hydrological history of the Sudd. ERTS and LANDSAT imagery will be examined to determine how much long-term monitoring of the system eventually can be based on these sources of information. The information collected will be used to develop an empirical model of the Sudd floodplain complex for use in making predictions about future events in the Sudd.

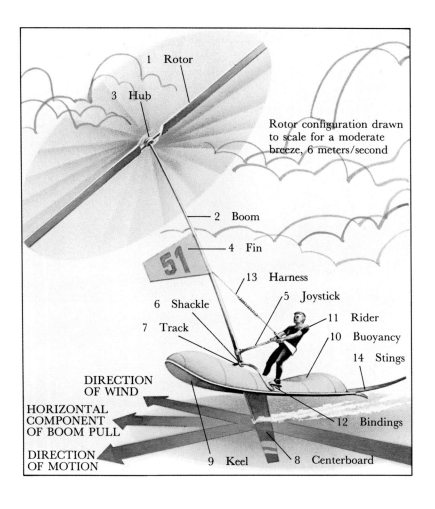

THE WHIRLYGIG — SAILING TO A NEW SPEED RECORD?

Since the ancient days of the triremes and before, people have sought to conquer winds and sea currents with crafts designed to go ever faster. Speed at sea became a question of combining brute power sources with appropriately adapted marine designs. Mathematical formulas invaded the realm of the boat- and ship-building craft. On-board computers have even taken their place as part of the equipment considered necessary to ocean-racing sailing ships. Little wonder, then, that most sailors believe modern wind-and-current craft have approached their speed limit.

But some believe that imaginative new approaches to "wind-powered vessels" will continue to break speed records. One such believer is an Englishman whose background in mathematics lends credence to his radically new sailing craft, which, if successful, may open a new vista on our timeless efforts to "fly" the seas.

NICHOLAS SHEPPARD
19, Bedhampton Road
Havant, Hampshire
England
Born October 2, 1951. British.

After leaving Trinity College, Cambridge University, where he received his M.A. in mathematics, Nicholas Sheppard worked for the British Antarctic Survey for nearly three years, mainly on the sub-Antarctic island of South Georgia. By the time he returned, he had managed to save several thousand pounds to support himself while he worked on two cherished projects: writing a book and developing an improved sailing craft, which eventually crystallized as the "Whirlygig."

I have invented a radically new type of sailing craft, the *whirlygig*, capable of traveling fast on water, ice, and snow, and of gliding independently. I have built a prototype, and my project is to perfect this craft, race it against other sailing craft (and if possible capture the World Sailing Speed Record), and otherwise promote the concept as a new sport.

The illustration of the whirlygig shows it with centerboard down, for traveling on water. The main component is the 10-meter rotor (1), spinning at the head of a slim, springy 7-meter boom (2). This rotor—exactly like an ordinary autogyro rotor—catches the wind and acts like a big disc-shaped sail or kite, keeping the boom always in tension. It is joined to the boom by a tiltable hub (3) on gimbals, again a standard autogyro fitting.

Halfway down the boom is a fin (4) to keep the boom and hub assembly properly aligned into the wind, and at the bottom is a joystick (5) that controls the tilting of the hub by a simple linkage.

The foot of the boom is fastened by a swivel shackle (6) to a sliding carriage on a curved track (7) like the mainsheet track of a yacht. This is arranged so the tension in the boom always acts through the middle of the centerboard (8), and thus there is no rolling movement. This inherent stability means that the whirlygig hull can be made much slimmer and lighter than an equivalent masted craft.

The keel (9) is like a 4-meter springy ski, with a Neoprene buoyancy tube (10) fastened along the top. The centerboard is attached to the keel through a spring linkage that allows it to hinge up and back when it meets an obstacle, or when the craft lands on snow or ice; then the task of preventing it from sliding sideways is taken over by runners and plates (not visible in the drawing) attached to the keel.

The rider (11) wears heavy boots like ski boots, which fit into step-in bindings (12) mounted above the keel assembly. Much of his or her weight is taken by a safety harness (13) anchored part of the way up the boom; this makes riding the whirlygig much less tiring than skiing.

The rider steers, on both land and water, by rolling the craft from side to side. At sea, this dips one or another of the flat, whippy "stings" (14) into the water and slews the craft around; on snow and ice, the sting cuts in somewhat, but most of the work is done by the keel and runners, which carve an arc, as a ski does.

Not shown in the drawing are other essential details such as the arrangements for keeping the handgrip of the joystick always comfortably within reach and for getting the rotor airborne in the first place. The essential principle is this: The rotor is supported only by its own aerodynamic lift. The boom is always in tension and only transmits this force to the groundcraft, like the string of a kite.

Naturally, this means that if the wind is too weak to enable the rotor to support its own weight, the craft cannot work at all. With a 20-kilogram rotor like my present one, this happens below a windspeed of 4 meters/second (m/s). When the windspeed is between 4 and $7\frac{1}{2}$ m/s, the rotor easily supports its own weight but cannot quite lift the groundcraft clear of the surface. It nevertheless gives a large sidewise pull, about the same as a conventional mast-and-sail arrangement.

Finally, with a windspeed greater than $7\frac{1}{2}$ m/s the rotor holds the groundcraft just clear of the surface, and the whirlygig, like the darting insect from which it takes its name, skims across the water—with no drag apart from the centerboard. Around the British Isles, the wind is strong enough for this effect about 45 percent of the time.

Because the rotor pivots freely in space, the rider has two degrees of control over it; he can vary the size of the force and can also distribute it at will between horizontal pull and vertical lift. When he comes to a large wave, he simply increases the lift component and flies up and over, instead of digging in, slowing down, and floating up, like a displacement yacht. On hard snow and ice, however, the lift must be eased off to allow the keel and runners to bite, except when leaping a crevasse.

As well as making short leaps at speed, the whirlygig can glide along clear of the ground, just like an ordinary autogyro. And the whirlygig can move more or less downwind over *any* surface that does not batter the keel too much—mud, reeds, sand or turf, for instance—and can lift itself almost vertically out of holes, bogs, and clinging seaweed.

At sea, the whirlygig can use the power of the wind itself to get out of trouble. Because of its speed over rough seas in strong winds, it can outrun the worst weather, and it can use the lift of the rotor to snatch itself from under a breaking wave or to fly itself over rocks or heavy surf to shelter ashore, where it is easily dismantled and stowed. Even in calm weather, it can be safely paddled far out to sea like a canoe with the rotor stowed—for if the wind blows up, so much the better.

The boom/rotor assembly can be unshackled from the groundcraft and used to move or lift dead weights or propel any other craft; it needs no special attachment point.

The whirlygig can race over quite large areas of sea, snow, and ice, with perhaps a few stretches of gliding, or slithering downwind over bogs, or slaloming among obstacles. It could be developed into a superb, satisfying sport. The present vogue for sailing, skiing, windsurfing, and hang-gliding suggests that it would soon become popular in a world where more and more people want to try their hands at wrestling with nature. The ultimate goal would be to have it accepted as a sport for the Olympic Games.

Development Plans

I have spent about 500 pounds on hardware, including a pair of Bensen Gyroglider rotor blades and a Campbell Cricket rotor hub. In many ways, it would have been more difficult to build a model autorotating rotor than a full-sized one, so I decided to go straight to the prototype's rotor.

I fitted this rotor to a special short boom, to test it for controllability, stability in gusts, vibration, and general handiness. The next step was fitting the complete control system to the short boom.

When I was sure I could control the rotor properly, I fit the full-length boom, perfected the mechanism for getting the rotor airborne at sea, and did some force and windspeed measurements.

Finally came the device for keeping the joystick within comfortable reach of the rider whatever the position of the boom, and then, complete with bindings and safety harness, I spent some time practicing with the finished boom/rotor assembly in different wind conditions.

All of these steps were accomplished with the boom-foot shackled to a block of concrete in a field. The next step is to fit the whole boom/rotor assembly to either a small boat (I own and sail a Mirror Dinghy) or to some kind of land-yacht, to get the feel of a moving craft. As I gain confidence, I will replace the dinghy with a canoe, do some speed trials, and finally fit the full craft out as in the drawing. At this stage, I will enter the John Player Sailing Speed Trials, at Weymouth, and any other events that offer the right kind of challenge.

Last of all will come experiments with free gliding (probably using an established hang-gliding hill) and with snow and ice—a bit difficult to find in England. I may have to go to Canada or Scandinavia to show off the whirlygig's capabilities properly.

My only problem is money. I may run out before I can finish the project. And this, of course, is where you come in.

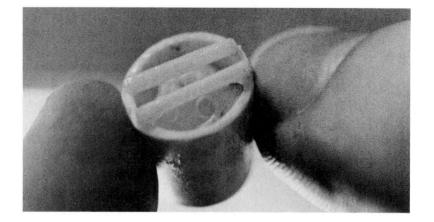

AIR POLLUTION — STOPPED AT THE SUBJECT, IF NOT AT THE SOURCE?

One of the hoary bits of advice to "problem solvers" is the old adage about there being more than one way to skin a cat. Occasionally, while others are busily attacking the most obvious side of the problem, someone looks at it from the other side and suggests a simple solution that had been overlooked. That may be what is happening in this project.

Although everyone agrees that airborne industrial pollution—especially particulate matter—is undesirable and even dangerous to breathe, the battle to force industry to clean up its waste air is slow and arduous, bringing scant solace to people who must work, live, and exist with the pollution. For these people the best quick solution may be to stop the danger where it strikes, at the personal level. And why not? We wear raincoats against the unstoppable rain; what about filters against unstoppable airborne pollution?

VIRGILIO ANDRÉS OLIVERA
Austria No. 2251, 30., A
Buenos Aires
Republic of Argentina
Born August 29, 1918. Argentine.

After a career in the Argentine army, from which he retired with the rank of major in 1971, Virgilio A. Olivera became interested in the problems of airborne pollution, particularly in industrial situations. He has been actively publicizing the solution that he proposes for people concerned about the air they breathe.

The main goal of my project is to make the nasal filter available to industrial and mine workers, as well as to the ordinary inhabitant of polluted streets. A secondary goal will be research into its use as a nasal container of medicines and as a respiratory protector for people with allergies. The project includes these steps: development of and research on the most convenient material for mass production, projection of the necessary molds and molding machines, research on the national and international markets, publicity in all possible media, series manufacture of the filters, and packaging and selling the product.

I patented the filter in the Argentine Republic on November 3, 1971, and in the United States on July 24, 1973. The patent represents a new technique of air respiratory filtration for humans: The filter is inserted directly into the nasal orifices instead of being placed over the face, as in masks or cloths. This nasal filter has been tested with my own handmade samples and has been scientifically certified.

The human body naturally protects itself through the respiratory system. Pollution of the environment, however, has recently become so extreme that the natural protective capacity of the body has been far exceeded. The industrial era, with its petrochemical processes, automobiles, pesticides, and the enormous amount of atmospheric pollutants, has therefore inspired this invention. The main idea is to clean the air before it enters the respiratory system and to allow free exhalation through the appliance without dislodging the filter.

Project Description

The invention is simple, but the solution took four years of work and problem synthesis. Unfortunately, distances and costs have inhibited full development of the invention and my ability to make it known in other countries—the best way to demonstrate its usefulness.

This invention is not a panacea; it will help, together with other weapons, to fight pollution. Its main characteristics are the following:

1. Its insertion is instantaneous, and its removal is simple.

2. Each filter must be lightly lubricated and totally sterilized.

3. One set of filters is needed for a day's work; if pollution is intense, two sets may be needed.

4. The housing is made of colorless plastic, the filter of synthetic fibers.

5. The device can be provided with a small projection to lock it in position in the nasal orifices or can be used without the projection if it is made in at least three sizes, to fit different sizes of nostrils.

6. The cost is low—the filter is cheaper than other appliances made to date, and, being disposable, it does not require any maintenance.

7. Its mechanical filtering effect blocks dust particles from 2.5 to 35 microns in size. The filters also may be provided with activated charcoal or other adequate products, to enhance its use as a chemical filter.

8. Another possible application of this filter is its use in inhaling medicines.

9. It is useful in mines, all types of factories, workshops, streets, and so on.
10. It should not be used by persons suffering from colds because breathing must overcome double the normal resistance.

11. It allows conversation during work and does not suffocate the wearer.

The project thus fights such problems as working absenteeism (35 percent of absenteeism in industrialized nations originates from respiratory illnesses), allergic illnesses (20 percent of which are due to respiratory intake of contaminant particles in the environment), and high smog levels.

We know that the human organism inhales about 8 liters of air per minute, which highlights the importance of breathing in healthy conditions and the necessity of raising a barrier between polluted air in the environment and the human lungs. To date, it is not possible to measure the effect on life expectancy caused by respiratory illnesses, but no doubt public opinion will soon force measures to drastically reduce this hazard. If people were asked to drink a polluted glass of water every day, they would refuse, but the same people may breathe in polluted air without having any control over its quality. Some day factories will have safe air and cars will not pollute streets. But when will that day come?

Applications

This invention has several applications, and the list can grow. The filter combats the following forms of pollution:

In Large Factories, Workshops, and Mines. Common dust, chemical products in the rubber industry, asbestos (lung cancer is characteristic of the asbestos industry), lead compounds, mercury, paint particles of several types, smoke emissions, gas and petrol residues, chemical products such as pesticides, coal and other dust dispersed in the air of mines, sulfur dioxide, nitrogen dioxide, and carbon monoxide.

In Small and Medium-Sized Industries. Pollution from factories producing DDT and other insecticides; spinning mills; iron and lead foundries; stone quarries; lime, kiln, and cement factories; coke foundries; fish powder plants; all other types of powder factories.

In Agriculture. Airborne pollen, pesticides, fungicides, herbicides, and so on.

In Streets and Other Places. Common dust, carbon dioxide and monoxide, smog, cigarette smoke, solid particles suspended in the air, lead from motor vehicles, and a long list of chemical compounds increasingly present in urban situations.

The nasal filter has many diverse applications, especially in countries with high industrial development and in large cities. This simple and inexpensive invention will be an efficient weapon in the battle against contaminants.

Parastichopus californicus, *the California sea cucumber.*

THE SEA CUCUMBER — A NEW FOOD

Through a combination of need, inventiveness, and economic realities, we have often succeeded in turning to alternatives from our established patterns in search of new benefits. New food products, such as peanuts and soybeans, have entered our diet because patient individuals pursued a belief that there is "a better way."

That same kind of inventiveness and enterprise may well serve our future needs as we explore the wealth of the seas. Apart from the huge fish and shellfish resources in the oceans of the world, what other food products in the underwater realm can offer us valuable and needed sustenance? In this project, we learn of someone who has followed his belief that he could find a new food in the sea and who has narrowed his search to a promising area.

LIONEL T. PENGSON
Honorable Mention, Rolex Awards for Enterprise
161 20th Avenue
Cubao, Quezon City
Philippines
Born January 14, 1946. Filipino.

After finishing primary school (1958), high school (1962), and business school (1965), all at Ateneo de Manila University, Lionel Pengson completed business school at Santa Clara University, Santa Clara, California, in 1967. He is president and proprietor of the Padagat Trading Company, which deals mainly in marine products, mostly seafood for export to Southeast Asian countries. As head of the company, he coordinates and manages almost all phases of the company's activities, from marketing to production to research. As a result, he has traveled extensively through Southeast Asia and the many islands of the Philippines.

The main purpose of this project is to discover if sea cucumbers can be cultured profitably. To do this, we shall have to explore and appraise the areas in the Philippines where these abound, the particular varieties indigenous to the areas, and the reasons for their abundance in a given area.

The sea cucumber, scientifically known as a holothurian in the phylum Echinoderma, belongs to a group of marine animals constituting one of the great branches of the animal kingdom. It belongs to the same family as the more well-known starfish. Shaped like a sausage, it comes in many different colors—brown, red, black, gray or white. Its length varies anywhere from 6 inches to as much as 32 inches. Approximately 500 known varieties exist in the world today.

In the Western world, the sea cucumber is more commonly known as *beche-de-mer*. The Chinese call it *hoi-sum*, the Japanese call it *namako*, and the Malays call it *trepang*. In the Philippines, it is known as *balat*. Although a great deal has been written about it, much more has to be learned for the world to benefit from it. The Chinese, who have eaten sea cucumbers for centuries, consider it to be not only a delicacy but also a medicine for treating high blood pressure. The reason for this may be its extremely high protein content.

In this day and age, when population is growing at an immense rate and food production is increasingly unable to keep pace, I believe it is time to look at the sea as a source of foods other than the traditional fish and shellfish that have been the most common seafoods since time immemorial. The sea cucumber, found as near as the tidal marks and as far as the greatest ocean depths, is one of the sturdiest animals in the marine world. It derives its food from organic materials found on the sea floor, and its metabolism rate depends on the availability of food. It is asexual, but certain varieties can break apart and grow the parts into whole individuals. This performance has been established as a regular mode of reproduction.

The Philippines is a country of more than 7000 islands. It has a total coastline of approximately 14,000 miles, slightly over that of the entire United States. Although our waters are very rich in fish, our food production from this area has been hampered by the lack of capital to purchase boats and equipment and by poor weather conditions, because our country is located in a typhoon belt that prevents year-round fishing.

In the past two years since we have started operations, our extensive travel through Southeast Asian countries and different islands of the Philippines has taught us some things about sea cucumbers. First, about 32 known varieties exist in Philippine waters. Of these, only four have commercial value in the Asian markets. Second, the right processing methods for each of the different varieties have been taught to the people where we have operated. We do this in the most primitive but efficient manner to help people living in rural coastal towns process sea cucumbers without special equipment. Third, although people from rural coastal towns know sea cucumbers exist in their areas, the majority have never really known what they are used for. Thus we have had to coin names for certain varieties. The problem of communication is compounded by the fact that dialects vary in different regions. Fourth, of the four varieties extant in our area, three grow in abundance in shallow waters not more than chest deep. Fifth, although habitats of different varieties do not differ much, certain varieties thrive more abundantly in certain areas. We have learned to pinpoint their areas on visual inspection. Sixth, among shallow-water varieties, the white sea cucumber has the most commercial value but requires the most tedious processing methods. However, the price it commands as compared with the other two varieties more than compensates for the additional effort.

The Project

In our proposed project, we hope to establish whether the white sea cucumber can be cultivated for commercial purposes. To do this, we must take certain steps:

1. Locate our research in an unpolluted area already inhabited by the white sea cucumber. We have done this already.

2. Secure government permission to exclusive operation in this area.

3. Fence off the area from intruders who might interfere with our controlled experiment.

4. Through the services of laboratories, measure and analyze the organic matter in the area.

5. Divide the area into several plots and make an inventory of the sea cucumber population for proper comparison after a period of one year.

6. Increase the organic matter, as determined by the laboratory, in half of the plots, while leaving the others alone. It is important here to remember that the metabolism of the sea cucumber increases with the availability of food.

7. Compare the facts and figures from the start of the project with those at the end and reach a conclusion. We will require a full-time marine biologist to record weather, organic input, sizes, population densities, and so on, during the entire project.

Benefits

We envision two major benefits that might accrue, should we prove the viability of this project and should we be able to enlist the aid of our government in our campaign to teach the people of coastal towns to cultivate white sea cucumbers.

First, such cultivation would increase earnings for people in rural coastal towns, who have traditionally depended only on fishing for income. Second, the people would have an additional source of protein food, which would alleviate malnutrition, a common problem in rural areas—especially during typhoon seasons when only rice or corn may be available.

RECYCLING OPHTHALMIC EQUIPMENT

Clinical ophthalmic instruments and equipment are costly items, a factor that pales when considering their key role in helping medical professionals to restore sight or prevent its loss. Because of the value we place on our sight, major research and development expenditures go into providing ever better, more versatile, and more sensitive pieces of equipment. Hence, "last year's model" tends to be replaced quickly in universities, clinics, and private practices of the industrialized world, giving way to newer and better devices.

What happens to last year's models? Gathering dust on a storeroom shelf, they serve no useful purpose even though their capabilities may not be in question. And what of all the students who came from less developed countries to learn the needed skills for working with the human eye? These students trained on many of the same outdated models, which are not available back home because they cost too much, when new, for the home country to purchase them.

There is an obvious conclusion. But for a long time it was not seen, before this sympathetic, understanding, and enterprising entrant put the two pieces together. It remains for the world to help.

PHILLIP HARRIS HENDRICKSON
Honorable Mention, Rolex Awards for Enterprise
Photodepartement
Universitäts-Augenklinik
Mittlere Strasse 91
CH-4056 Basel
Switzerland
Born November 6, 1943. American.

After receiving his bachelor of engineering degree from Vanderbilt University, Nashville, Tennessee, in 1965, Phillip H. Hendrickson went on in 1967 to earn his master of science degree in sanitary and public health engineering from the same university. At the time of submitting his project, he was an ophthalmic researcher and photographer and instructor of ophthalmic photography in the department of ophthalmology, Joban Municipal Hospital, in Iwaki City, Japan. Since that time, he has moved to Switzerland to take up new responsibilities in the Basel University Eye Clinic.

The "Rolex Award Project—Opportunity Recycling in Ophthalmology" (RAPORO) is to be a pilot project for recycling used, but still usable, clinical ophthalmic instruments and equipment from industrialized countries to developing lands. The activities of RAPORO will include contacting potential sources of such ophthalmic hardware in institutions and in industry, determining the transport means, and selecting and supplying the recipient countries.

During my ten-year career in the field of ophthalmic research, I have had the opportunity to live and work in the United States, West Germany, and Japan. Moreover, this experience has also offered me a chance to become acquainted with doctors and technicians from developing Third World countries (DC's), who most often have had to carry their modern training and skills gained in industrialized countries (IC's) back to their homelands, where they are severely and discouragingly handicapped by the lack of all but the most primitive instruments and materials. Furthermore, having lived in IC's, I have become aware of the abundance of "replaced" equipment (that is, used but still usable), which, because it is assumed to be worthless and obsolete, has been relegated to storerooms, back shelves, and attics of many institutions and industries.

Currently, DC's are asking for (even demanding) a global redistribution of wealth. I believe that a realistic, immediately achievable redistribution of opportunity could be embodied in the recycling of so-called obsolete equipment from IC's to DC's; therefore, I wish to start and conduct a pilot project for such material recycling, limited to my field of ophthalmology, under the project name of RAPORO. From this proposition arise several points that must be clarified: (1) defining the relativity of *obsolescence* in IC's, (2) establishing sources of such equipment within IC's, (3) transporting the equipment from IC's to DC's, and (4) determining needs and priorities of DC's.

1. We in the IC's are well aware of the dramatically rapid advances in technology in our countries, both in goods and in services. And improved communications have made us better acquainted with the fact that only a relatively small percentage of the world's population can enjoy such benefits. The rest lag behind, indeed advancing, but at a much slower rate. However, rapid advances in technology in IC's inherently cause a rapid technological obsolescence. In DC's, the much slower rate of development, as well as the much slower rate of distribution of goods and services, means that, by lagging behind IC's, DC's still regard the already obsolete technology of IC's as advanced, relative to that of their own countries. In the course of eventual redistribution of global wealth, it should thus be possible, even preferable, to proceed to help countries modernize DC's by way of the intermediate step to be explored by RAPORO.

2. Establishing sources of recyclable equipment by necessity will involve the cooperation of as many consumer institutions as possible. Contact would be made with professional ophthalmological societies in IC's and, through the society structures, with each individual ophthalmologist or clinic, which would appraise its own ability to contribute to the recycling effort. Furthermore, the cooperation of supplying industries (for example, optical companies that supply instrumentation for ophthalmology) would be solicited to establish RAPORO as a potential user of trade-in equipment and to determine the availability of equipment for recycling. Tax deductions and the chance to contribute to a humanitarian project should encourage institutions and industry, which can generate good public relations as well as future marketing possibilities. Contact will also be made with the XXIIIe Concilium Ophthalmologicum Universale to be held in Kyoto, Japan, in May 1978, thus affording a chance to acquaint ophthalmologists worldwide with RAPORO. RAPORO's proposed starting date of November 1, 1977,

will provide sufficient time to prepare preliminary organization and consultation before that congress meets in May 1978.

3. Procuring transport of recyclable material from IC's to DC's requires an availability study of possible help from the shipping industry, which will be requested to place unfilled space at the disposal of RAPORO. Tax and public relations encouragements will be offered in this area, too. A certain portion of the Rolex Award will be, by necessity, channeled into shipping costs for smaller items that can be transported by parcel post. RAPORO will endeavor to obtain preferential postal rates because of RAPORO's humanitarian nature.

4. Finally, determining the specific needs and priority rankings of target DC's will necessitate consultation with the appropriate organizations. On September 1, 1977, I will begin my employment in the Universitäts-Augenklinik in Basel, Switzerland; I consider this country, as location of the headquarters of the World Health Organization (WHO) and the International Red Cross, to be an ideal base of operations for RAPORO. For example, I plan to ask WHO's Study Group on the Prevention of Blindness for help in targeting, soliciting, and screening potential recipients of RAPORO equipment. In this regard, I will also ask the XXIIIe Concilium Ophthalmologicum Universale for initial consultation and publicity for RAPORO. In general, it is anticipated that one target country on each continent be initially chosen to test the recycling mechanism. The number of target countries and locations then can be expanded in proportion to the supply of recyclable materials.

I have naturally considered the possibility that insufficient material turnover could occur in the course of RAPORO's expected two-year duration. However, RAPORO is conceived as an investigation into the feasibility of such recycling.

A DRUG PROBLEM OR A CULTURE PROBLEM?

The mere suggestion that what we Westerners refer to as a drug problem might conceivably be something other than just that is an invitation to ridicule, anger, and other forms of rejection. Yet, despite massive efforts and major expenditures to combat the drug problem, we still seem to be far from comprehending the fundamental causes of the problem. It is too facile to suggest that the increased availability of the drugs in our midst has created the problem. We must look deeper into ourselves for the reasons why drugs, over the last 20 years or so, have become more available. Have we truly nothing to learn from cultures that have used drugs for centuries and yet have no social phenomenon they consider a drug problem?

The project described here grapples with these questions in methodical and logical fashion. Its results should be of interest to all who are concerned and are coping with an uncomfortable and growing reality—the increased use of drugs.

FLORIAN DELTGEN
Honorable Mention, Rolex Awards for Enterprise
Erftstrasse 93A
5159 Sindorf
West Germany
Born September 3, 1940. German.

Florian Deltgen's bilingual background—his father spoke French, and his mother spoke German—probably influenced his later interest in languages (he has learned English, Latin, Italian, Russian, and "some" Greek, Spanish, Ki-Swahili, and the Bushman click-language known as !Koe). These proved to be of great help to him in his studies in ethnology at the universities of Freiburg, Paris, and Köln (1962-69), prior to his receiving his doctorate (summa cum laude) from the University of Köln in 1969.

His work as a wissenschaftlicher Assistent *(assistant professor) at the Institute of Ethnology of the University of Köln includes teaching, research, administrative management of the institute (shared with his colleagues), and counseling students.*

His general area of study is the ethnology of Latin America, with emphasis on the Indians of Colombia's tropical rainforest. Within this field, his main interests are ethnopharmacology, the socialization of children, ecstatic religious phenomena, and methodology. He spent two years preparing for the field work that he describes here.

The project is field research among a group of Arawak-speaking Curripaco Indians of the Colombian tropical rainforest. I want to collect information about the process of socialization of male youths in general and to learn how the Indians supposedly structure individual drug experience within a specific group setting.

The idea that in order to communicate with supernatural beings one must alter one's state of mind is found in almost every religious system. A variety of altered states of mind has been labeled *ecstasy*. The consumption of strong hallucinogenic drugs is only one of many ways to produce ecstasy.

Since the early 1960s, the use of hitherto unknown or unusual drugs has been steadily increasing in all industrialized Western countries. Not only is the type of drugs taken new, but so is the type of drug takers: teenagers and preteenagers. This situation has created much conflict and enormous problems, about which much writing has consequently been generated. Most of these books and articles identify a drug problem, suggesting that the drugs, as material substances, are regarded as the source of the trouble. In essence, they say that there is a drug problem because there are drugs.

My survey of ethnographic literature about drug use among nonliterate peoples has strongly suggested that this view is wrong. If, for example, the pharmacologic properties of dimethyltrypamin (DMT) *alone* were responsible for a variety of negative consequences, these consequences should be the same everywhere. However, they are not. The average effects triggered by the consumption of this drug differ greatly from culture to culture. Even among cultures of the tropical rainforest we can observe differences, although the drug remains the same. It seems the difficulty is not so much a *drug* problem as it is a *culture* problem.

The survey also revealed that ethnographic sources do not provide the information needed to analyze and solve this problem. All I could do was to establish some hypothetical assumptions. First, we must distinguish between a formal and a contents component in individual drug experience; only the formal component is determined by the pharmacologic properties of the drug administered, while the content of the experience is determined by the culture to which the individual belongs. Second, we must assume a polarity between individuals and groups. If, in a given society, there is a great deal of instruction about the use and the effects of drugs, about how one reacts and behaves under their influence, and about what people experience, a high degree of standardization for individual experience will result. Such instruction will also integrate the drug complex into the social and spiritual life of the group. But if there is little or no instruction at all, the individual experience will be predominantly structured by one's psychic "set" or attitude. In this case, we can expect poor integration of the drug complex into the social and spiritual life of the group and, as a consequence, a high degree of conflict. Third, standardization of the psychic and physical reaction of individuals toward the formal effects of a hallucinogenic drug can be achieved by means of teaching techniques and learning processes; that is, by some sort of drug pedagogy. Fourth, if, through the study of a less complex society, we could find out how drug use functions in principle, we might obtain a model that could help us to cope with the problem of drug use in a way very different from those we have already tried.

I knew that if I wanted to provide more, and more relevant, information about this problem, I would have to do field work. I decided to do it and designed the project as follows.

Problem

The problem was defined by the following questions. Can the content of experiences originated by the pharmacologic properties of certain hallucinogenic drugs be effectively standardized and, if so, by what means and measures? What is the effect of successful standardization: more integration of the drug complex into the culture of the group and less conflict, or vice versa? More stable personalities and interpersonal relations, or less stable ones?

Empirical Design

I knew that we must find a nonliterate group with a simple social structure. This group must have a long tradition in the use of an hallucinogenic drug. It must have a relatively well-preserved traditional cultural identity; that is, it must not have undergone any long-lasting acculturation, especially not any missionary activity.

Such a group is no longer easy to find. It was clear from the beginning that only the South American tropical rainforest could provide an ethnic group meeting our requirements. We chose the Curripaco Indians of Colombia for the following reasons: first, we do not know very much about them; second, they live in an extremely remote area and were reported to resent any contact with whites; third, one of the few things we know about them is that they have a custom of drinking yagé, a beverage made from *Banisteriopsis caapi* and/or *Virola* spp., both of which contain the hallucinogen DMT (dimethyltrypamin).

The Curripaco, who belong to the Arawak linguistic family, live between the Isana and Guainía Rivers, along the lower Río Guainía, and near the lower Río Inírida in the vicinity of the Mesa de la Lindosa in eastern Colombia, Comisaría del Guainía. The heart of their country is $1°30'-2°30'$N latitude and $68°-70°$W longitude.

Now we must get there, establish friendly relations with the Curripaco Indians, learn the language, and collect data about which elements of the culture are connected with the drug, what the Curripaco do with the drug and what it does to them, and whether the individual is left alone with the effect of the drug or whether the experience is influenced by other persons (and if so, by whom and how)? We must ask Indians about their experiences and compare their reports. And we must observe directly those processes by which the expected standardization is actually achieved. Also, we must try to find out whether persons under the influence of the drug show significantly more deviant, aggressive, or unsanctioned behavior than persons in a normal state of mind.

I worked out a research plan and submitted it to the West German Federal Board for Scientific Research, who decided to finance the project. All preparations are now made. I shall be accompanied by one German and one Colombian, as coinvestigators. Our time schedule is as follows:

December 1976	Explore area from the air and try to get to a place where a group of Curripaco can be spotted.
January 1977	Start a three-month stay with the group, during which the only aim will be to learn their language as soon as possible.
April–November 1977	Conduct explicit observations in order to provide data concerning drug use.
December 1977	Return to Bogota and then to Köln.

Our main problems in the technical realization of the expedition are energy supply, storage of films and tapes, transportation, maintenance of health, and communication. We will take an electric generator, hunting equipment, waterproof plastic barrels, a 25-horsepower outboard motor, medical instruments and medication, and a five-band, 500-W shortwave transceiver. We know that the whole enterprise is extremely risky because of the isolated area and the unusual length of the field job.

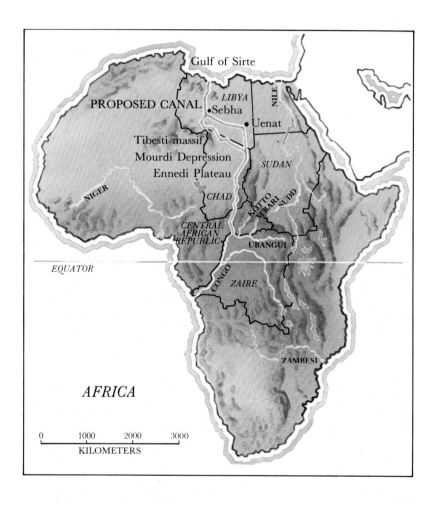

A NEW GREAT RIVER FOR THE WORLD?

The romance and mystery of the great rivers have always fascinated people. The Nile, the Amazon, the Mississippi, the Yellow, and many others—all suggest some special bond between humans and nature in a way that few other natural phenomena do. The great rivers are almost tangibly alive, and they have lent that life spirit to the civilizations that have clustered along their sustaining banks. As sources of food, facilitators of agriculture, trade routes, and, in times of flooding, terrible destroyers of the mere mortals too near their paths, the great rivers have demanded and received our close attention and curiosity.

The searches for their origins comprise one of the most exotic chapters in the annals of exploration, one virtually closed for want of further material, for these mysteries of nature have long since succumbed to the explorers' curiosity.

But suppose that there remains one last great river story, an exploration that would cap all those previously recorded. Would that strike your fancy? Come, we are going to Africa. . . .

ROBERT LIVINGSTON POMEROY
10 Lathbury Road
Oxford
England
Born September 26, 1915. American.

As a retired US foreign service officer and former newspaperman (with the New York Herald Tribune *and* United Press International*), one might expect Robert L. Pomeroy to have seen a fair portion of the world. In fact, beginning with his early schooling, he lived in China, Italy, and Switzerland before attending universities in the United States, where he studied English literature, European history, geology, biology, ornithology, and engineering. He has traveled through the Sahara as far as the Tibesti Mountains, but never as far south as the area he would be exploring in this project.*

I propose a survey and feasibility study for the construction of a freshwater canal from the Ubangui River, northward through the Central African Republic, Chad, and Libya, to the Mediterranean.

Between the Congo Basin and the Mediterranean, there is enough fertile land to provide a good living for millions of people and produce food for millions more. All that is needed is enough water, a means of distributing it, and a way to carry the produce of this land to markets of the world. In other words, what is needed is a river—a new, great river. The fundamental question is, could we ever build such a river?

The project I propose attempts to determine whether it would be geographically and economically feasible to survey and construct such a waterway. A project of this sort raises many questions, of which the most basic are the following:

1. What would be the advantages of building such a waterway?

2. Where would the water come from, and is there enough of it?

3. Is there enough altitude to carry this water for 3000 kilometers from its source to the Mediterranean?

4. Would the final cost of construction make its study worthwhile?

5. Finally, what would a preliminary study entail in personnel, time, and cost?

Taking these questions in sequence, I would answer them as follows.

First, a waterway large enough for both irrigation and navigation in this part of the world would do far more than bring hope for a better life to the people of the countries through which it passed. As the waters would be drawn from the rivers of Zaire and the Central African Republic, both of these countries would thus earn some revenue. They would also benefit by flood control and would receive a certain amount of electric power from the system of dams that would be built to divert and impound these waters. A ship canal through this part of the world not only would be useful for transporting livestock and other agricultural products to market but would also make it possible to develop and ship mineral reserves that cannot be moved economically today. Finally, it would make available thousands of square kilometers of unpopulated lands for new settlers, and eventually it could even bring tourism to this beautiful land.

Second, the source and adequacy of the water are one of the aims of the study. Perhaps all of the requirements could be met by the Ubangui and its tributaries. There certainly would be far more water than needed from the Congo, which is fed from a year-round rainy season. The question is how, rather than where, it would be gathered. What dams would be required; what tunnels, pipelines, reservoirs, and pumping stations would have to be built; and where should they be placed?

Third, the Ubangui would undoubtedly be used as one of the principal sources, with the main storage reservoir on one of its tributaries fairly high up on the sides of the Mangos massif—perhaps along the gorges of the Kotto or M'bari rivers. This collection basin would have to be high enough (between 500 and 600 meters) to allow a continuous flow to the sea. Starting at 600 meters and flowing for 3000 kilometers, the average fall would be 1 meter every 5000 meters. Although this would not result in a fast-flowing river, it would keep moving, and traffic could move up or down without difficulty. The Nile falls only about 400 meters from the Sudd to the sea, which is also about 3000 kilometers away.

The greatest problem would be mapping the course this waterway would have to follow. Obviously, it could not drop into the Chad Plains but would have to skirt along the foothills of the northern slopes of the Mangos massif, paralleling the Sudanese border, and continue this general northerly direction at least to the Ennedi Plateau. At some point, the route would probably veer northeast almost to the Sudanese border, to avoid the Mourdi Depression, which becomes a deep canyon at its western end. North and west of this lies an area of about 250,000 square kilometers, much of which has never been explored and certainly not surveyed, except from the air. The mean elevation here is about 400 meters, and although there are a few ranges of low hills the greater part of the area is hard, flat sand and gravel pan. Only an accurate survey could determine the best route from here.

Either it would veer westward, cutting through the plains north of the Tibesti foothills and then northwest into the Fezzan, or it would continue almost due north to Uenat, where it would turn northwest to pass about 16 kilometers south of the Kufra Depression and on westward into the Fezzan to join the route first suggested, at or near Sebha. If both routes proved possible, eventually they might both be used, bringing that much more land into use. Whichever of these routes is taken, from Sebha the course would undoubtedly go almost due north, through the pass in the Djebel Uadan, to empty into the Gulf of Sirte.

Fourth, the cost could be determined only after the feasibility study and survey had been completed. It is obvious that a construction of this scale, which might rank with humanity's greatest achievements, would cost in the billions of dollars. Even so, it would be less than many countries spend on arms every year. Perhaps, for political or economic reasons, this waterway will not be built—at least not in our lifetime—but the study and survey are within reach and would show future generations that there were people today who thought about tomorrow.

Fifth, and last, the field work probably could be completed in a year, but unfortunately both Zaire and the southern Central African Republic have a long rainy season, and the Sahara becomes almost unbearably hot from May to September. This means that the actual field work would probably extend over two years. The summers could be used for preparing maps and reports. As far as personnel are concerned, there might be as many as ten. These would include a hydrologist, geologist, surveyor, navigator, doctor, pilot, and radio operator, but one would hope to find some who combined one or more of these specializations. The cost of the entire project would be somewhere between $100,000 and $200,000. This figure would depend on how many of the technicians came as volunteers rather than on salary and on how much of the equipment could be rented rather than purchased.

In reviewing this project, one might reasonably ask, "If this land is worth providing with a system of transportation and irrigation, why not use a railroad or automobile road, flanked by a pipeline for water? Wouldn't these be less expensive to build and to operate?" Yes, they would be less expensive to build if one is speaking of a small—say, 30 inch or 1 meter—pipe, but far more expensive if one tried to design a pipeline large enough to do the job that an open ditch can do. The proposed waterway would cost more to build but less to maintain and use, and consequently it could deliver both water and freight at a fraction of the cost of any other combination of systems. There would be the problem of water loss by evaporation, but this would be true of any river flowing through a hot, dry land. In theory, it might be argued that the waters of the Nile could never reach the sea, but of course they do.

Probably the best argument for this or any other river is what the early Egyptians thought of the Nile: It represented the roadway of life.

THAT PARAPLEGICS MAY WALK AGAIN

Few tragedies in the world are more heartrending than those accidents that, although in other respects reparable, cause the partial or near total severing of the spinal column, leaving an otherwise intact and vital human being consigned to virtual immobility for the rest of his or her days. Medical science has long known the reasons for the resulting paraplegia (paralysis of a limb), but apart from providing care, correct diet, and physical therapy little has been possible to arrest the gradual deterioration of the muscles that leaves the patient permanently incapacitated.

Now, thanks to the determination and perseverance of a young French medical student, there is real hope that these patients may once again be able to walk, assisted by a brilliant merging of medical scientific knowledge with space-age technology. The tools are now available, and the goal is clear. Time and money should bring success in the not-too-distant future.

GEORGES MARCEL ANDRÉ DELAMARE
Rolex Laureate, Rolex Awards for Enterprise
6, Avenue Général Leclerc
Boite Postale 59
54500 Vandoeuvre
France
Born December 13, 1953. French.

A young medical student who hopes to earn his doctor of medicine degree in 1979, Georges M. A. Delamare is the youngest of the Rolex Laureates. Dividing his student time between classroom and hospital experience, he works on a heavy schedule to achieve the degree he needs in order to continue his efforts to make the scientific breakthrough he describes in his project.

This project is a new rehabilitation method for paralyzed, mainly paraplegic, people. The method entails computerized electronic stimulation of nerves and muscles to generate motion and walking in paraplegics.

Injuries to the spinal cord after a traumatic lesion cut the connections between the muscles and the brain and leave the patient with a great handicap, such as paraplegia (in cases of horizontal sectioning of the spinal cord) or paralysis of a lower limb. Often patients with quadriplegia must spend the rest of their lives in bed.

The target of this project is to help the paraplegic to walk with his or her own legs and muscles, assisted by electrostimulation synchronized by computer.

Basic Principles

When a motor nerve in a muscle is stimulated by an electric pulse, a contraction occurs. The strength of the contraction depends on three factors of electrostimulation: duration, intensity, frequency. The threshold of duration and intensity vary with the type of muscle, and in human muscles a 50-hertz frequency determines the maximal contraction, called the *tetanic contraction*. If two electrodes are placed on the muscle, the electric activity of the muscle, amplified and recorded, gives an electromyographic study.

It now appears that electrostimulation can be performed on humans and that an important part of contractile strength can thus be saved. W. Steinberger has shown that maintenance of denervated muscles allows the patient to keep 35 percent of contractile strength. Steinberger states that "If no more than a third of normal tension-generating capacity could be preserved in human denervated muscle over the long term, this level would still be useful in powering orthosis systems." Contractile strength can be increased through appropriate diet (proteins), nerve stimulation (in paraplegia, motor conduction velocity of nerves is normal), and electrostimulation soon after injury.

Description of Project Details

The lower limb is divided roughly in two main groups of muscles: *extensors* on the anterior part and *flexors* on the posterior side, plus adductors and abductors on the interior and exterior sides. The upper extremity is composed of four extensors and three flexors, and the lower has four extensors and eight flexors.

Electrodes are implanted on every muscle, which is not a surgical problem, although the electrodes must be chosen to prevent infection and breakage. The electrodes are connected to a frequency generator and to a power source (battery). A detector sends information about the patient's vertical position to a computer, which instantly analyzes spatial position and determines the correct response; that is, the right frequency and intensity to apply on extensors or flexors to keep the patient upright if he or she is bending backward or forward (pressure detectors placed under the feet also provide information). This measurement and correction are now possible because a computer can integrate all the data and immediately produce a response in agreement with experiments performed previously to determine a muscle's correct contractile strength.

For the patient to make a step, the computer must synchronize contractions between the different muscles of the same group because they do not contract at the same time. A program for the walking sequence is in the computer's memory and is used to apply electrostimulation in agreement with all the data received from the patient's position and muscular capabilities.

The main problem is not to synchronize the two legs (only one is necessary because it is a symmetric movement) but to prevent the patient from falling. The patient now represents a dynamic system, able to start, stop, and turn left or right. To prevent stumbling it is necessary to place accurate accelerometers on the patient that will inform the computer of variations in the gravity center. Electrodes for electromyography are placed on the patient's neck and control walking direction. If the patient turns his or her head to the right side, the contractions of the neck muscles are recorded and sent to the computer, where, once analyzed, electrostimulation is sent through the abductors.

This method can be improved by microcomputers placed on the patient; these eliminate the stationary computer and the need for telecommunications between computer and patient. For the time being, patients must command their movements by pressing buttons. This action may be needless in the future if the analysis of cortical potentials becomes more accurate, in which case movement could be directly determined through brain command.

Paraplegic patients could walk. The data have been known for several years, and the appliances have existed for ten years. In fact, the first step of the paraplegic patient will be a great step for humanity and a felicitous merger of technology and medicine.

DID THE AZTECS USE WOOD FOR BONE IMPLANT MATERIAL — AND CAN WE?

Sometimes we simply overlook the obvious until some insightful thinker points to it. Such may well be the case with the obvious questions raised by remnants of skeletons and skulls found in the Aztec burial grounds that show serious damage not necessarily related to the cause of death. "Missing" sections of bone and neatly trimmed holes in skulls raise the question: "How did life go on in spite of these apparent damages?"

It may be that the missing bone material was not replaced by other bone material but by particularly well-suited wood fittings that have long since rotted away, leaving little evidence for nonmedical archaeologists. There seems to be little doubt today that under certain circumstances, wood can be used as an effective bone implant material. In this intriguing quest, the search for answers may lead to a trail hundreds of years old.

HERBERT KRISTEN
Honorable Mention, Rolex Awards for Enterprise
Reisnerstrasse 48
A-1030 Vienna
Austria
Born June 19, 1939. Austrian.

An assistant at the Orthopedic Clinic of the University of Vienna, Dr. Herbert Kristen took his M.D. in orthopedic surgery in 1963. He has co-published a paper on the subject of his project and has lectured on the subject of experimental surgery.

Certain wood species show physical and elastic properties similar to those of bone tissue. For the first time, our research could prove the possibility of bonding bone tissue to wood. Wood could become quite important as bone implant material.

While continuing our experimental research work, we plan, partly depending on the outcome of this application, to search for suitable tropical wood species, according to certain selection criteria, that may improve the results we have achieved already with local varieties. We also plan to do direct research, with archaeological field teams in Central America (especially Mexico), on skeletons that may contain Aztec wood implantations.

One of the greatest problems in orthopedics is the search for implant materials that can replace bone sections after operations on sarcoma and metastasis. These materials must have appropriate strength properties, ease of adjustment and fitting qualities, elastic behavior, compatibility, and so forth.

Locally grown wood species, such as ash (*Fraxinus* sp.) and birch (*Betula* sp.), which have been thoroughly studied in our basic research project, offer easy access to the necessarily large amount of preselected specimens. The compatibility, strength, and elastic properties (especially the latter) seem to be satisfactory. The porous structure of these woods is another advantage. It is quite possible, however, that certain tropical wood species might already, or after treatments, have an even better combination of the required criteria.

To introduce these species to research, much basic work and screening can be and are being done with small specimens shipped by mail through existing contacts with forest products research institutes in tropical countries and in Europe. But to select larger usable implant sections it is absolutely necessary to have direct access to trees (or to specially cut and treated boards) in the country of origin, plus nondestructive testing of preselected pieces of wood and a second, more careful selection in Austria.

Through intensive literature study, we have found only one reference suggesting that Aztecs used wood as implant materials in bones. A scientific description of skeletons, and especially skulls (trepanations were also closed with hardwoods), could not be found. It is quite possible that the full implications of long-bone fractures and of openings supported or closed by wood were not considered in examining the remnants or relics because the wood had rotted away over the centuries.

For our basic project, it would be of great interest and impact to undertake this research work in relevant museums in Central America, Mexico, and elsewhere. Discoveries of wood implants in Aztec skeletons and skulls would encourage our original work in implant surgery and would verify the high standard of Aztec medical science.

Our current plan is to continue the experimental tests with overseas wood species after preselection and screening. In the summer of 1978, we would travel to Central America and conduct experiments with animals for ten months. Evaluation of this anatomical research could be completed by the end of 1978, assuming that animals for experiments and materials for histological research would be at our disposal. Research concerning the strength and elastic properties of the woods would be done within the facilities of the Institute of Wood Research in Vienna.

The following is a list of tropical wood species that might, because of their physical properties, be suitable for implant material, although some of them contain certain highly toxic substances that must be extracted before use: Afrormosia, Balata, Bethabara, Blackbutt, Celtis-African, Curupay, Danta, Ekki-Bongossi, Eng, Esia, Greenheart, Jacareuba, Jarrah, Kapur, Karri, Kempas, Keruing, Kurokai, Lancewood, Lignum vitae, Manbarklak, Mansonia, Marish, Mengkulang, Mora, Muhimbi, East African olive, Omu, Padauk, Central American pine, Purpleheart, Sterculia, Sucupira, Tallowwood, Tasmanian myrtle, Tasmanian oak, Tatabu, Tonka, Wacapou, Wallaba, Wamara, and Wattle.

Canis lupus, *the gray wolf.*

COMMUNITY SURVIVAL IN THE ARCTIC — LESSONS FROM PEOPLE AND ANIMALS

One of the more interesting observations of our human condition, made by sociologists, is our tendency to group into communities numbering between 70 and 80 people. Studies of primitive village societies and of personal phone book listings of city residents suggest that we tend to form contact groups up to about that number and then level off.

How much this grouping is required for survival purposes is not clear, but it occurs in all social animals, of which we are only one species. One view suggests that different species in the same harsh environment adapt in similar fashion not only to the requirements of survival in their surroundings but also to each other as fellow combatants in the fight against an unyielding nature.

In the Arctic, small communities of people and animals have evolved a way of life that has kept them in balance with nature and each other for untold centuries. This project proposes to learn more about that relationship, in order to help us understand our own, less clearly defined ability to survive as members of a complex society.

JOHN CARROLL FENTRESS
Armdale Rural Route 5
Box 30, Site 12
Halifax County, Nova Scotia
Canada
Born February 4, 1939. American.

The project that John C. Fentress describes here quite logically combines the two hats he wears in his regular working life. First, he is a psychology professor (and chairman of the department) at the Life Sciences Center of Dalhousie University. Second, he is director of the Wolf Research Project at Shubenacadie Wildlife Park, Nova Scotia. His academic background includes a B.A. (1961) from Amherst College in Massachusetts (summa cum laude in psychology and the Woods and Travis Prizes for outstanding student) and a Ph.D. (1965) from Cambridge University, England (in zoology, specializing in ethology, supported by a National Science Foundation fellowship).

He went on to become a postdoctoral research fellow at the University of Rochester, where he worked on neurobiology and behavior, on a US Public Health Service fellowship. He then became an assistant and an associate professor in the departments of biology and psychology at the University of Oregon, where he founded and directed the BioSocial Research Center and directed the behavioral biology training program. He has been at Dalhousie since 1974.

In the harshness of the Arctic, wolves and Eskimos have each survived, not by banding together in large numbers nor by living solitary lives but by forming small cooperative groups. However, a new threat now endangers the survival of both—modern technology and our insatiable need for resources that are limited. These aspects of our society often destroy social and biological environments on which human and animal lives depend. This project has been designed with an awareness of the speed at which we are encroaching on the Arctic and of the lessons still to be learned about the wolves and the Eskimos, about their relationships with each other and their environment, and, ultimately, about ourselves.

Both wolves and Eskimos live in small cooperative groups in one of the few relatively undisturbed environments on this planet, the Arctic, which is now threatened by our expanding population, technology, and demand for rapidly diminishing resources.

For the past decade, my wife and I have examined social order in captive wolves and have published the results of our research. I am a zoologist and behavioral scientist; my wife is a naturalist, professional photographer, and student of diverse cultures. We see the wolf as a creature of beauty in its own right, as well as a wild animal whose life involves social cooperation, family groups, and biological adaptability. A study of wolves can yield both scientific insights and poetic inspiration.

My wife and I have worked very intimately with wolves in captivity. We have hand reared and socialized several animals, observed others undisturbed in large wooded enclosures, probed the subtleties of their communication, and extensively filmed and taped a variety of the dynamic relations that define their behavior.

Our goal in this project is to extend our work in two directions. First, we plan to observe wolves over a period of months on Baffin Island, in the Canadian Arctic. Second, we plan to communicate with the Eskimos who share this region with the wolves.

As recently discussed by Nicholas Tinbergen, Nobel Laureate in ethology, an understanding of convergent rules of adaptation among diverse species who share a common habitat can yield important insights into the roots of our own existence. Within this context of important tasks for ethology, Tinbergen explicitly mentions wolves and humans living in populations that are "small, scattered, and dwindling fast." As a professionally trained ethologist, I agree with Tinbergen and wish to participate in this important enterprise.

Our view contains a second mission. A clear lesson of art, science, and culture is that we to a large extent construct our own realities. That is, we fashion our world; we do not merely observe it. The Eskimo has fashioned a world of harmony with nature, a sense of continuity between human and animal. We expect to learn much from the way in which Eskimos view wolves within their own world. By studying animals and people together, we anticipate insights that would otherwise be inaccessible.

We hope to spend about six months exploring Baffin Island, observing its wolves, and communicating with its native inhabitants. Our emphasis in this study will be on the relationships of cooperative society in animals and humans.

To aid us in this venture, we plan to bring with us a highly socialized male wolf (now one year old). This animal can help us locate other wolves and generate an appropriate focus for our communications with the native peoples. The animal will be outfitted with a collar that houses both a transmitter and receiver, to facilitate constant communications. A full documentation of our explorations will be made, including still and movie films, tape recordings, and detailed notes. Further, we have developed grammatical computer programs that permit stochastic (temporal) analyses of the deep structure in expressed behavior for animals as well as humans.

We have a Land Rover at our disposal and plan to use it both for transport and for housing. Most basic photographic and recording equipment and other electronic devices are also available. Funds are needed primarily for transport, supplies, small equipment items, and living expenses.

The project is planned to last from April 20, 1978 to October 20, 1978. This period will permit close observations of wolves during their denning season.

One important focus of our study will be to explore contextual determinants of adaptive behavior. What are the similarities and differences in social order between free-ranging and captive wolves, for example? To what extent does behavioral convergence in humans and animals reflect their environmental contexts? How do the various features of adaptive behavior relate to one another as a functional whole?

Within this framework, we plan to focus on problems of patterned behavior as hierarchically ordered expressions through time, where both ephemeral and more stable qualities of behavior are viewed together. These studies can lead to relational or relativistic models of adaptive behavior in temporal as well as spatial boundaries. We have evidence for the importance of temporal features in both the perception and the behavior of wolves, and it is equally evident that diverse human cultures view temporal connectivity in nature in different ways. In many important respects, time is behavior, because behavior through time is a critical component of adaptive relations between living organisms and their environments.

To summarize, our ultimate goal is to provide a broad biological assessment of adaptive boundaries in behavior, including temporal, quantitative, and qualitative characteristics. We feel that a careful analysis of wolves and Eskimos and their relations to one another in the Arctic environment can provide a relativistic perspective on biological adaptation at various levels. Not understanding these issues, as Tinbergen notes, can lead to loss of viability for both animals and humans. The issues also have implications for building a dynamic systems approach to biology (as mirrored, for example, in current theoretical physics).

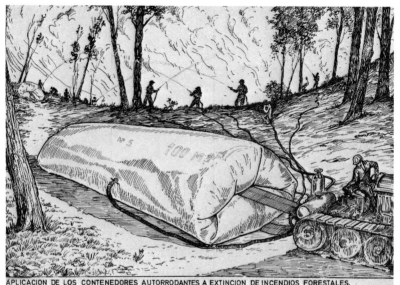
APLICACION DE LOS CONTENEDORES AUTORRODANTES A EXTINCION DE INCENDIOS FORESTALES.

EMERGENCY WATER IN HUGE "SELF-ROLLING" BAGS

Be it forest fires threatening homes high up in canyons, or drought conditions that dry up a critical water hole, or temporary needs for water at a distant construction site, the demands for massive amounts of water in locations with no local supply are frequent. The inability to provide water quickly and easily poses problems ranging from the economically frustrating to the catastrophic.

Assuredly, we can arrange to transport water to many desired locations by installing permanent—and expensive—systems of canals, irrigation ditches, and even pump-supported pipelines, but such answers fail to meet the need for *temporary* supplies.

In an experimental effort underway in Spain, an intriguing means of temporary water supply is being investigated on a pilot scale. Organized at an industrial level, it could provide many parts of the world with answers to local water problems.

FRANCISCO ALCALDE PECERO
Santa Clara, No. 40
Sevilla, 2
Spain
Born July 18, 1941. Spanish.

After receiving his degree as an architect technician from the University of Seville in 1967, Francisco Alcalde Pecero went on to become a professor in the Seville School of Technician Architects, where he teaches architectonic details and does research and investigation.

I propose studies of a new system that transports large volumes of water in self-rolling containers hauled by conventional tractors. This system could rapidly supply water on unforeseen occasions for agriculture, livestock, fire extinguishing, and other uses.

Water, as we know, is a vital raw material, and the manner in which we get it, store it, or contain it may help to develop a country. Until a few years ago, its availability depended mainly on an adequate supply and how it was stored. This kind of provision is inadequate for current rapid development; it has not met the needs of certain circumstances and has even had catastrophic effects on certain underdeveloped countries.

The problem is a pressing one, under study in all countries, because the progressive increase in population, with resultant increases in water consumption by people, industry, agriculture, and others, will result in water shortages for agriculture and livestock farms in certain zones. International organizations are trying to find new sources of supply by developing new techniques, such as desalinizing seawater by electric power, finding and using subterranean water, producing artificial rain, collecting dew water, and particularly using all surface water by avoiding its contamination.

We recognize that some zones have more abundant water resources than others, and that this varies according to climate. For reasonable water use, the problem consists of transporting water from abundant zones to ones having water shortages. The present means of transportation require costly channeling, irrigation ditches using gravity, expensive concrete or metal piping with complicated pumping systems, or large aqueducts of varying continuous flows. These systems are costly, and their design and construction take a long time. They do not solve emergencies caused by drought unless their construction has been completed, and they do not help if water has to be transported to faraway zones, except in very favorable circumstances.

Other means of supplying or transporting water—until now little considered—could be the intermittent supply of large volumes of water as needed, which could offer considerable advantages. A great technological gap exists here, and no sufficiently rapid, cheap, and efficient system has yet been developed.

People with avant-garde ideas might imagine complicated and enormous water-tank vehicles of rigid shape and constant volume for sporadic transportation, which could roll (on wheels or caterpillar treads) or slide (on air mattresses). However, if we think carefully, we can see that such transport would require enormous loads and roads capable of bearing such loads.

The object of this project is to study a new system for the occasional transportation of great volumes of water by means of self-rolling, flexible, all-terrain containers, capable of being towed.

The system is constructed on a new and simple geometric form, rotating an element like the caterpillar of a tank (chain treads) around a longitudinal axle that coincides with one of its straight sides. When a small force is applied to its center, this container can move forward by rolling in the direction of its length. To start it moving, we put an endless strip through its center to remain anchored by the interior hydrostatic pressure caused by the container's being full of water; the pressure makes the strip solidary with the nucleus of inside folds in such a way that when a pulling (or traction) force on the strip is transmitted to the center, it produces forward movement. If this endless strip is hooked to a pulley system of free rotation and applied to a conventional tractor, when the tractor moves the whole thing will move.

Manufacturing the container is simple. It is made of a gummed textile in the form of a pipe, twice the length of the final container, with valves installed. One end of the pipe is joined to the other one by glue or vulcanization, creating a leakproof chamber. When filled with water, the container is subject to hydrostatic pressures from all directions in proportion to the amount of fluid put into it. The superficial tensions produced are absorbed by the textile material and its vulcanization (which keeps the material hermetic).

The endless strip that tows the container with pulleys applied to a conventional tractor is a strip of synthetic material resistant to traction. Its width depends on the inside pressure of the container, the power of the tractor, and the angle of the grade to be climbed. It could also be a cable or chain with transverse filaments.

The project basically consists of studying the container itself, materials for its construction, valves, strips, pulley systems, volumes, diameter-length relations, durability, behavior on various types of terrain and over itineraries such as lands with grades longitudinal and transverse to the container, as well as gearing for changes of direction within a minimum radius.

Our goal with these studies is to obtain containers with capacities of more than 100 cubic meters (m^3), tow speeds of 10 km/hour, distributed pressure over the land less than that of the human foot when walking (0.3 kg/cm^2), a relation of tare-to-load of approximately 1/100, a cost of support) to measure the effects of the traction
$0.01/$m^3$/km (amortizations and transportation expenses included), and capability of climbing 15 percent grades.

The project comprises numerous elements:
• General studies.
• Built-in valve designs.
• Simulated experiment on a testing table: Roll a small container of approximately 1000 kg of water over an endless strip on rollers, inclinable in two perpendicular directions to simulate upgrades and equipped with control and measuring apparatus:

1. Dynamometer fixed to the pulleys (which will be attached to a fixed support) to measure the effects of the traction

2. Revolutions-per-minute meter synchronized with the pulleys to control speed of advancing

3. Revolution meter to control the distance run and to initiate resistance tests of the gummed textile to be used

4. Apparatus speed changer that will move the endless strip

• Contacts with manufacturers of gummed textiles to select the most appropriate fabric.
• Based on data obtained from the experimental table, a 25-metric-ton device and its equipment will be designed and manufactured.
• Study and design of a field trial on a closed course approximately 500 meters in length with obstacles, such as curves and upgrades, of the sort encountered on a real itinerary, plus some controlling apparatus.
• In accordance with the results obtained during field trial with the 25-metric-ton container, study, design, and manufacture a container of 100 metric tons and improve container by experience obtained from the smaller container.

- Trial of the 100-metric-ton container on the closed course.
- General conclusions, practical studies, documentation, printing of manuals, and so on.
- Preliminary studies have practically commenced on a small scale, with our own means. The entire project depends on the availability of further funds.

The advantages that can be derived from this project, we think, are important:

- Rapid supply of fresh water for irrigation or for industrial use to zones that for unforeseen reasons may need it.
- Transportation of water cross country, covering 100-km distances in 10–12 hours.
- Lightweight equipment, easily transported by any means, even by air (the 100-metric-ton container plus equipment, without the tractor, weighs about 800 kg).
- Supply temporarily inhabited camps or supply desertlike zones in periods of acute drought.
- To stock forests with several full containers strategically located, to be used in case of fire and capable of being transported quickly over long distances.
- Small containers of up to 25 m^3 to supply water to cattle concentrations and to carry water inside industrial installations.
- The future possibility of carrying much greater volumes of water in containers of more than 100 m^3.
- Transportation of other liquids, evacuation of contaminated water or other liquids to appropriate places, and so forth.

The specific aims of the project are to be able to transport water in containers of 100 m^3 to places located at a distance of over 50 km from the source and to solve the problem of using this system in rugged terrain.

We are optimistic about the project. It would provide a simple and definitive method for transporting water in certain important need situations, and it would give world conservation agencies a cheap, rapid, and efficient means of combating drought, thereby helping not only people but also animals faced with the threat of extinction. It could also irrigate great desertlike zones with water from faraway places. I would turn over these techniques to the public.

There is no risk of total failure in our investigations, in our opinion, because the system is simple and clear and because experiments have already been carried out with these containers in the construction of concrete channels.

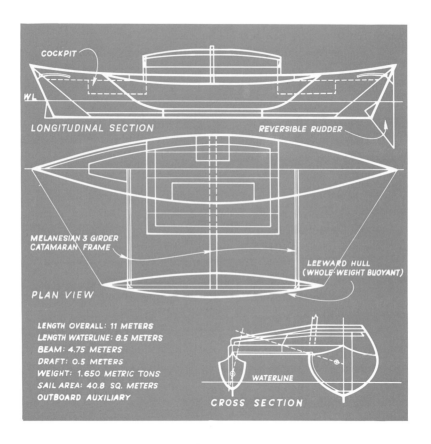

RETURNING TO THE OPEN SEAS IN MODERN PIROGUES

For many primitive societies, the impact of modern technology can have both beneficial and tragic consequences. In the first blush of "newer and better," a shift away from traditional skills and undertakings means that a certain craft or expertise atrophies in a few generations. The loss is sometimes apparent only later, when the new technology has been assimilated and fitted into existing patterns.

Such a case can be made for the great ocean-sailing traditions of the South Pacific, in which coastal tribes roamed over vast expanses of sea to fish, trade, explore, and even socialize. After centuries of experimentation, sailing craft of unusually appropriate design became common throughout the area, despite rigorous construction requirements. A "new era," with readily available motor craft, has put a virtual end to the building of ancient pirogues; along with that, the water territories covered have become drastically limited. The resulting concentration of coastal population and activity has polluted lagoons and severely affected the wide-ranging fishing industry.

Today, with the ability to simplify the intelligent boat designs of the past, there may be a chance to restore ocean-sailing craft as a much-needed part of local economies.

BERNARD ANDRÉ FRANÇOIS DUJARDIN
Honorable Mention, Rolex Awards for Enterprise
Boite Postal 431
168 Anse Vata
Noumea
New Caledonia
Born January 5, 1940. French.

Following his graduation in 1962 as an ingenieur de l'ecole navale from the French Naval College and his 1963 graduation from the Ecole d'Application des Enseignes de Vaisseau, Bernard A. F. Dujardin went on to the Institut des Sciences Politiques of Paris, where he graduated from the public utility program in 1969. This was followed by his graduation from the Ecole Nationale d'Administration in 1972. This combined naval and public governmental background undoubtedly serves him well, both in his position as vice-secretary general for economy of the French overseas territory of New Caledonia and in the project he proposes here.

I plan to develop a project for Pacific islanders that is based on a traditional Melanesian boat design, the pirogue, built with modern materials and European small-craft technology. The objective is to revive the Melanesian pirogue—not for folklore preservation (the ancient technology is too good to vanish) but for economic purposes: the development of energy-saving bonito fishing activities within the craft industries of coastal tribes.

I believe that ancient Melanesian seafaring craft would be safer, faster, more convenient, and more versatile than European ones on a dimension-for-dimension basis (weight, sail area, and length on water line) *if* the original designs were brought up to date and executed with present-day technology. During the time that Melanesian people have been in touch with Europeans they have come to prefer small European utility craft over the Melanesian pirogue, which is difficult to build in its traditional dugout form. Since they have begun to sail these imported boats, they no longer put out for the open seas because they quite rightly consider them inadequate. The ocean sailing tradition of these people has sunk into oblivion. Fishing areas are limited in coastal lagoon waters, where disease restricts fish consumption to certain species and times and forbids any real development of an export trade.

Bonito fishing vessels must avoid the lagoons to reach their fishing grounds. In New Caledonia, this means a transit course of 15 to 20 nautical miles beyond the barrier reef. As compared with the four to five hours needed to make a round trip of this type with a 7–8-knot diesel-powered craft, a wind-on-beam pirogue (the normal weather conditions in the trade belt, where New Caledonia lies in the wind's eye) would cut the traveling time to two or three hours. Fishing time would increase by 20 percent, and costly fuel would be spared. These advantages are made possible by the astonishingly steady trade wind encountered locally (12–18 knots during the daytime).

As of now, there is no bonito fishing economy in New Caledonia. Such a development requires a native crew and the adaptation of a suitable vessel. Ultimately, it means building, sailing, and testing the boat, and showing the Melanesians its ability by an open-sea crossing.

As shown in the drawing, the craft is asymmetric longitudinally and symmetric athwartship. It is a decked Melanesian (North New Caledonia) pirogue with an outrigger. It tacks either by reversing the direction of travel (the outrigger kept leeward in heavy seas and weather) or in the usual way (sailed with the outrigger windward in fair weather). The outrigger float can sustain the full weight of the craft. This type of craft appears to offer the ultimate in speed and stability as compared with any monohull, catamaran, or trimaran sailing vessel.

Rudders and sail tacks are reversible in order to substitute the bow for the stern, and vice versa. The draft of longitudinal keels on both hulls can be distributed for the best balance. That is, balancing the effort of sails on both tacks keeps the draft as shallow as possible—a safety factor in the coral seas. The mast is stepped at an angle to the plumb line and tilted windward as in ancient designs, thereby keeping air flow on the sails free of disturbances and allowing simple handling of the rig. Construction is based on the traditional three-girder structure and double longitudinal catamaran frame design. This structure is strong, can gently dampen the waves, and is easy to build.

Hull material will be cork and glass-reinforced plastic or plywood, for which the design should be redrafted with a two-chine section. The choice will be made according to local facilities. The decks and bulkheads and general arrangements will be built of plywood. Beams and mast will be made of aluminum, the running rigging of polyester rope, the standing rigging of stainless steel, and the sails of polyester.

Once finances are available, four to six months would be needed to build the craft in a yard, depending on the availability of supplies. I allow three months for trials, and then the craft could be sailed on a demonstration crossing. After this, we will experiment with the boat as a utility and fishing vessel. If she performs well, we hope to create a production line in a tribal boatyard.

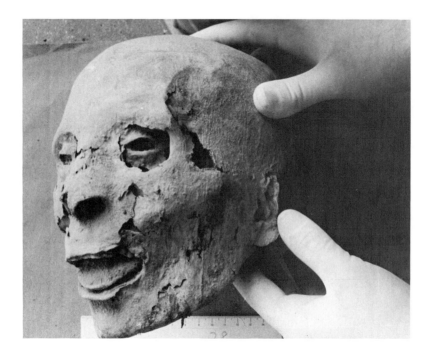

HAVE WE INVENTED CANCER?

The day may come, in the not too distant future, when an inspired combination of hard work, scientific creativity, and solid research leads to a cure for cancer. Even then, however, we may fall short of medicine's ultimate victory—the ability to prevent the disease—unless we can ascertain its causes.

A growing body of evidence suggests that cancer's increase is somehow linked to our own civilization—that we have somehow created the disease ourselves. Proof of this hypothesis could significantly narrow our search for the key causes.

This project proposes to establish a point in human history when cancer may not have existed. To fix such a time with strong confidence would help us discover what we ourselves have done to bring this disease upon us.

MICHAEL RAYMOND ZIMMERMAN
Department of Pathology
University of Michigan
Ann Arbor, Michigan 48109
United States of America
Born December 26, 1937. American.

After earning his B.A. in 1959 from Washington and Jefferson College, Michael R. Zimmerman went on to get his M.D. in 1963 from New York University's School of Medicine. In 1976 he received a Ph.D. in anthropology from the University of Pennsylvania, where he continued as an assistant professor of pathology and anthropology. In this position, his responsibilities included surgical pathology, teaching, and research in paleopathology. He subsequently moved to the University of Michigan.

This project proposes to expand a preliminary study of Egyptian mummies to allow a statistically valid statement regarding the incidence of cancer in ancient Egypt. A University of Pennsylvania expedition has removed a large number of mummified bodies from a New Kingdom tomb (ca. 1300 B.C.). Thirty of these mummies already have been examined; no evidence of cancer has been found. Studies of approximately 20 other Egyptian mummies and of Peruvian and Alaskan mummies have yielded similar results. Experimental mummification studies indicate that the paucity of tumors is not due to poor preservation, and there is ample evidence that in antiquity many individuals lived to advanced age. The virtual absence of malignant tumors in these limited studies encourages further work and suggests that the current increase in cancer is due to environmental factors in the modern world.

Research Plan

The objective of the proposed study is the development of a historical perspective on the incidence of cancer. The etiological implications of the apparently low incidence of malignancies in antiquity will be considered in detail.

The rationale of the study is that the rehydration of mummified tissue by a water, alcohol, and sodium carbonate solution allows the preparation of microscopic slides and the diagnosis of virtually the entire range of pathological conditions, including neoplasms.

Background. Mummies are defined as bodies preserved either naturally, as by drying or freezing, or artificially. The Egyptian practice of artificial mummification developed from the natural preservation of bodies buried in the hot, dry sands of the desert in pre-Dynastic times. At the beginning of the Dynastic period, above-ground tombs made it necessary to develop artificial techniques of mummification.

Until the Christian era—about A.D. 209–400, all deceased Egyptians were mummified. Techniques varied throughout the millenia and with the wealth and social status of the deceased. Not until New Kingdom times (the period of the mummies of the proposed study) was the body thoroughly desiccated by the heat of sun or fire, after a period of maceration in a salt solution. Thus, these later mummies are generally well preserved and suitable for study.

There has been a recent resurgence of interest in the examination of mummies. An earlier peak had been reached with Sir M. S. Ruffer's work in pre-World War I Egypt. Ruffer developed a rehydration technique that remains in use, in its original or modified form, and also introduced into paleopathology the use of microscopy.

After Ruffer's death, little systematic work was done on Egyptian mummies until the recent formation of the Paleopathology Association. This group of scientists is pursuing multidisciplinary investigations of mummies.

It is noteworthy that there have been only two tissue diagnoses of neoplasms in Egyptian mummies. Sandison reported a small, benign, squamous papilloma of the hand in a late Dynastic female, and Zimmerman diagnosed a dermatofibroma (also benign) of the skin of the heel in a mummy fragment. (Urteaga and Pack did make tentative diagnoses of malignant melanoma in Peruvian mummies.) Although gross diagnoses of malignant tumors of various types have been made in skeletal material and in the skeletons of mummies, there has never been a microscopic diagnosis of a malignant tumor in ancient Egyptian material. This includes the 30 Egyptian mummies that I examined in the preliminary study.

Although in antiquity the average life span was short, many individuals did live to the "cancer age," as there is ample evidence for a variety of degenerative disorders, including degenerative joint diseases, atherosclerosis, and Paget's disease of bone. It has been suggested that tumors would not be well-enough preserved for diagnosis, but tumors that have been experimentally mummified and rehydrated are actually better preserved than normal tissues. Thus, the absence of tumors in Egyptian mummies studied to date must be considered a reflection of a markedly lower incidence than in modern industrial populations, in which cancer affects approximately 25 percent of all adults.

Specific Aims. The specific aim of the study proposed is the expansion of a preliminary study of mummified material from Dra Abu el-Naga, Egypt, including dissection and microscopic examination. We plan to examine systematically the remains of up to several hundred individuals. Should the rate of malignancy remain at or near zero, a definitive statement on the evolution of the disease will be in order.

Procedure. The procedure consists of the sorting and gross examination of the fragmented mummified remains. The specimens are photographed, dissected, and selected for gross (macroscopic) or potential microscopic pathology. After clearance by the Egyptian Board of Health, the specimens are shipped to the laboratory at the University of Pennsylvania for rehydration and microscopic examination. As far as is possible with material of this nature, standard microscopic descriptions and final diagnoses are written, and appropriate correlations are made with archaeological data.

The hazards of working with mummified material are chiefly those of inhalation of dust or possibly microorganisms. The investigators wear surgical masks and gloves at all times when working with the materials, and shower facilities are available in the nearby living accommodations. Conditions in the geographic area, 400 miles south of Cairo, require a full regimen of immunizations.

The tentative schedule is for the field work to be performed in March 1978. The specimens should arrive in Philadelphia by November 1978, and $1\frac{1}{2}$ years will be allowed for examination, interpretation, and publication of the material.

Significance. The apparent absence of malignancy in ancient Egypt has significant implications for the etiology of the disease. It is difficult to find accurate statistics for cancer incidence in modern Egypt. It is estimated that 95 percent of the fellaheen (peasants), who make up

80 percent of the population, are afflicted with schistosomiasis. This parasitic infestation predisposes to bladder cancer, which is relatively common in modern Egypt. Failure to find this tumor in an adequate sample of ancient Egyptians, among whom the prevalence of schistosomiasis has been demonstrated, would suggest that some additional carcinogen, necessary for cancer development, is present in the modern environment. The absence of other tumors would carry similar implications and strengthen the view that up to 75 percent of human cancers is related to environment.

Facilities. The facilities available include standard photographic and dissecting equipment. The dissections will be carried out in a convenient field office, and the microscopic sections will be prepared in the Histopathology Laboratory at the hospital of the University of Pennsylvania, which has full facilities for standard and special strains, as well as microscopy and photomicrography.

ENGINEERING THE BALANCE BETWEEN MAN, LAND, AND WATER

There is no shortage of concern among scientists for the growing "desertification" of vast areas in arid and semiarid regions of the world. Many of these regions were once fertile, as one can see from the traces of river networks and from the ancient records of their populations. Water was key to the fertility, and the loss of the water may well have been accelerated by the demands of an early agricultural system that did not account for the preciousness of a commodity vital to the long-range future of these early societies.

Many of today's efforts to combat the growing deserts have concentrated on further tinkerings with the environment—dams, reservoirs, irrigation canals, and the like. It may be that we have overlooked the ability of the land itself to reassert an earlier climate through the growth of proper vegetation. In this project, with the aid of a newly perfected scientific instrument, there may be the opportunity to engineer a more productive future by incorporating previously unavailable knowledge in the search for solutions.

JAROSLAV BALEK
Honorable Mention, Rolex Awards for Enterprise
Kopeckého 8
16900 Prague 6
Czechoslovakia
Born April 26, 1933. Czech.

Prior to becoming senior researcher and consultant at Stavebni Geologie Corporation for engineering, geological, and hydrogeological investigations, Jaroslav Balek was senior researcher at the Institute of Hydrodynamics, Academy of Science, Prague. He took his civil engineering diploma in water resources management, irrigation, and drainage from the Technical University, Prague, in 1956. Earning his Ph.D. in hydrology and water management from the Academy of Science, Prague, in 1963, he went on to become a postdoctoral fellow with the National Council of Scientific Research of Canada in 1967 and attended a special course in environmental engineering at the University of Moscow in 1975.

Environmental protection is one of the most discussed topics of our time; however, so far we have no definition of an "optimal environment." Despite rapid technical development, in many parts of the world the population suffers from periodic, or permanent, water shortages. We are aware of basic facts; for example, a standard of living equal to that of Europe cannot be achieved by normal means in semiarid and arid regions. What *should* be accomplished in these areas, at least, is to balance nature, climate, and human beings as much as possible.

Climate is not changeable through any reasonable economic means. Our own way of life is difficult to readapt; in fact, it is, and will continue to be, a permanently disturbing factor in the environment. In nature, only water resources and vegetation are foreseen as suitable for regulation. Nevertheless, water resources are managed mostly by unnatural means, such as the construction of dams, reservoirs, and canals, which, besides being costly, have negative side effects and frequently disturb the environment. Water resources also have certain limits of exploitation, owing to available rainfall and groundwater storage factors. Thus only vegetation, as a main consumer of water, remains a factor that can be regulated.

The effect of vegetation is most pronounced in the marginal parts of the deserts, where it forms a natural barrier against the desert extension. Until now, however, very little has been known of the hydrologic water balance and its variability in these areas. In particular, the role of vegetation has not been fully explored.

Theoretically, with under 100 millimeters (mm) of rainfall per annum, less than 1 meter of soil becomes moist, thereby allowing only the shallow roots of grass to grow. About 300 mm of rainfall is necessary to provide conditions favorable for shrub and tree growth. Even at 500 mm, conditions are still dry, and the vegetation protects itself with thin leaves, thick bark, and deep roots. Such a state of equilibrium is found rarely under natural conditions because, in addition to rainfall, soil moisture and groundwater formation play equally important roles. Besides these natural factors, the role of humans must be accounted for in semiarid areas; in an attempt to cultivate the soil, people frequently burned vegetation.

The remains of river networks in arid and semiarid regions indicate that these areas were developed before the present stage of aridity and desiccation occurred. The resulting obliteration of the lower parts of the river networks produced a vegetation that changed from an intermittent and perhaps perennial variety to one of ephemeral growth. Thus the present stage of these networks represents an adjustment to the environment and indicates that aridity is not a constant.

Observations of arid and semiarid regions confirm that a substantial increase in water yield occurs after brush vegetation is replaced by grass, because some species can evaporate more than twice as much water as barren soil can. (Between 190 and 500 mm of water can be saved in a year in arid regions by removal of phreatophytes.) Yet with the disappearance of the woody plants or even reduction of their height, the microclimates deteriorate, and, later, so do the macroclimates of semiarid regions. As the action of wind on plants and soils becomes more marked, temperature variation on and below the soil surfaces increases, infiltration is affected, and evaporation becomes more intensive. This also explains why trees and shrubs, once cleared, no longer reproduce. All these problems figure in the effectiveness of desert fringe vegetation in acting as a protective barrier between the desert and subhumid regions and as a natural regulator in water management.

The following questions summarize the problems: (1) What are the chances for, and the constraints on, the regulation of the hydrologic cycle in regions with sparse localized water supply? (2) What are the means of effective regulation, and what are their limits? (3) How can conclusions be generalized beyond the experimental limits?

Operational Plan

We propose to organize an intensive search in one or more semiarid regions or desert margins, such as in the Sahel, for vegetation cover suitable as a hydrologic and ecologic regulator and also acceptable to human life. In contrast both to the standard hydrological approach, which is based on water balance calculations, and to descriptively involved ecological methods, we propose to concentrate on the study of transpiration as the single most significant water balance component.

We intend to gauge vegetation transpiration by accurately measuring the sap stream velocity of the species growing inside or near the boundaries of the region. The experiment should be carried out with both temporary and continuous measurements. The temporary measurements should be made in the field by a group, consisting of a hydrologist, an ecologist, and an electronic engineer, that would travel through the region. Continuous measurements should be made on the transpiration of selected species different in type, age, and location. Because sap stream variability is a matter of minutes and seconds, extremely accurate timing methods for the entire experiment will be essential.

The appropriate instrument for such measurements has already been developed and tested on a small scale in experimental areas in Czechoslovakia, and preliminary results were published in the *Journal of Hydrology*. Some minor improvements in the instrument are foreseen in advance of the experiment.

Basically, the project idea resulted from my study of land use in afforested areas of the Central African Plateau. The project would be conducted during the dry periods of two years (probably 1978 and 1979), according to the region selected for the experiment.

EVALUATING THE ROLE OF PREVENTIVE MEDICINE IN AFRICA

Much progress has been made in Africa in combating the ravages of killing and debilitating diseases. More and more clinics and trained professionals are joining the fight against illnesses that affect sizable portions of the African populations. Public health systems are increasingly able to provide preventive medical care—the important next stage of the medical battle.

In the project described here, a dedicated Congolese tells of the success he and his team were able to achieve in a program to prevent the worst illnesses in children, who are highly vulnerable. The results are good, and the system is spreading, because the people themselves see its benefits demonstrated in dramatic personal terms.

JEAN-RAYMOND MISSONGO
Honorable Mention, Rolex Awards for Enterprise
77 Rue Madingou (Moungali)
Brazzaville
Congo
Born December 12, 1946. Congolese.

Jean-Raymond Missongo is presently studying medical and social law in France, to acquire the knowledge he needs to continue the project he describes here.

This concerns a program of preventive medicine in the Congo. The World Health Organization (WHO), recognizing the importance of this plan, which sets an example for all the surrounding countries in our region, has assigned it two expert advisors. However, the really basic and important work is carried out by Congolese national health officers who, in spite of enormous difficulties, have already made great strides in integrating preventive medicine with curative medicine programs. This is the only health policy capable of bearing fruit in a region of the world that suffers the classic scourges of measles, malaria, tuberculosis, leprosy, and trypanosomiasis.

First I will briefly review the essential characteristics of the situation and will define the limits of the project. The principal target of the project is the Pool Region, chosen for its homogeneous structure and population and because a rural development plan run by the International Work Bureau already exists there. The plan, which has been published by WHO, will introduce the new health policy in a region where the population already has a standard of living more tolerable than elsewhere in the nation. The population (close to 200,000) represents the highest density in the country (nearly 15 inhabitants per square kilometer). It can be divided up, unequally, as follows:

1. Age 0–2 years. The mortality rate for infants (because of malaria) is more than 15 percent in a hyperendemic zone such as the Congo.

2. Age 2–5 years. The mortality rate of this 10 percent of the population attains a level as high as 50 percent, which neutralizes the high birth rate often attributed to our countries. The causes of death are malaria (responsible for 10–15 percent of all deaths), parasitic diseases, and, above all, measles and tuberculosis in children already weakened

by imbalanced diets (including a premature deprivation of mother's milk).

3. Children of School Age (6–15 years). Having resisted the attacks of the environment, these "survivors," who constitute 25 percent of the population, will acquire a certain immunity to disease.

4. Adults. Sanitary protection must be given to adults (80–95 percent are healthy carriers of hematophylic parasites) in order to tame these diseases.

5. The Elderly. Old people, who are often neglected, frequently constitute a dangerous reservoir of Koch bacilli, which can infect younger members of their families.

The project, which focuses on the principal regional health center, can be subdivided into some five sections; the three most important are hygiene, social needs, and epidemiology. The last especially concerns us here, for the policy of prevention and prophylaxis in this part of Africa depends on epidemiology. I direct this part of the total plan.

My team is using two methods to integrate preventive and curative medicine. First, in the maternity and infant clinics, we have introduced antitetanus vaccination for future mothers, in two stages (the sixth month of pregnancy and before confinement). As a result, the mortality rate from neonatal tetanus has diminished considerably. Second, in all the health training centers in the principal zone, we have introduced compulsory nivaquinization of babies and children 0–5 years old. Weekly intake of nivaquine (50 milligrams for infants 0–1 years old, 100 mg for children 1–5 years old) has reduced the number of deaths from the most dangerous form of malaria, that caused by *Plasmodium falciparum*, which is unique to the Congo. Before this preventive measure was introduced, this plasmodium caused deaths and destruction that were attributed to taboos, sorcerers, or fate; these beliefs have declined as the people realize that malaria is a medical problem.

Our ambition is to protect the children 0–5 years of age, who previously suffered a terrifying death rate (50 percent). The acceptance of chemical prophylaxis has been tremendous. Public authorities, together with village officials, form a village health committee that designates an officer responsible for the weekly nivaquine distribution. The people soon understand the importance of this strategy when the children no longer have convulsions and nocturnal fevers. In Kindamba, one of the four districts where this system has been most effectively used, the people publicly denounce the distributor to the local authorities if he fails to carry out a single nivaquine distribution. The nivaquinization of children is taken very seriously throughout the whole country as the result of the experience gained through "Project Pool."

We studied absenteeism of children attending first-year preparatory classes (ages 5-6), and, at the request of the majority of primary school teachers assigned for periods of study in the health field, we have already introduced nivaquinization of schoolchildren, based on experience gained in the schools of the regional centers of Kinkala, BoKo, Mindouli, and Mayama. This program is only operative during the school year, that is, from October to June. Nevertheless, it has greatly reduced the rate of absenteeism, previously estimated at close to 25 percent, according to a survey I personally carried out at Yokama and Matoumbou.

Much has been done to aid the heads of the health centers to integrate preventive medicine with health training, by stressing periodic sensitivity to predictable diseases, such as tuberculosis. As a result, two methods of discovering disease have been put into practice. One is the practice of bacilloscopy for all patients after two weeks of nonspecific treatment. Additionally, the bacilloscopic procedures of Ziehl-Neelsen and Lapeissonnie-Causse have been used widely in the larger centers, paid for by a WHO gift of several thousand Congolese francs for laboratory work.

Since September 1973, we have divided the ordinary work of the epidemiological section into two parts: programs against specific diseases and programs for prospecting and vaccinations.

The specific programs are specialized campaigns against major epidemiological problems. In the antipolio program, vaccination is carried out annually for children aged 3 months to 6 years. We have used the vaccine trivalent, which originated in China and which has a form (three candies) acceptable to children. Children in all regions have been vaccinated, with the exception of one region whose very sandy roads require a "go-anywhere" vehicle that was not available. The results of this campaign have definitely freed us to concentrate on the simultaneous struggle against many diseases.

Next is the onchocercosis campaign. Until now, we have limited ourselves to sounding the alarm with regard to this dangerous filariosis. Originally, we found cases only in BoKo (on the banks of the Congo), but on our last trip through the Mindouli District, in June 1975, we discovered at least five other cases, finding *Onchocerca volvulus* by directly puncturing the nodules. This area, bordering on Zaire, is very different from the BoKo District. We had to limit our strategy here to the distribution of a diethyl-carbamazine (notezine) treatment that only reduces the microfilariemnine population and thus delays the fearful end result of the disease, namely blindness. More than 100 blind carriers of nodules and even of microfilaires have been discovered in the BoKo District. All this constitutes a gross public health problem, demanding the united action of volunteers from world organizations (as for the Upper Volta project).

The leprosy problem is more delicate, for lepers need more medical care. Each district has three officers responsible for distributing disulone. During our leprosy control tours, we distribute various food products (rice, fish, meat, milk, and cheese) donated by the World Food Program or by the national leprosy committee. Also, the international benefactor Raoul Follereau, whose name has become legendary in our country, gave us several transport vehicles. We suggested the creation of a center for the most contagious lepers (in the lepromatose stage), for preliminary treatment only. This center is now in operation at Brazzaville.

The second aspect of the epidemiological work, the prospecting and vaccination tour programs, is not yet in full operation. We need to add a complementary team of five officers from Operational Sector 1 at Brazzaville, as well as supplies for this particular strategy. We need at least two "go-anywhere" vehicles of the Saviem type: one for materials (gas freezer for vaccines, work tables, microscopes, beds, and individual eating utensils) and one for personnel. Completing the whole campaign in the principal zone takes at least eight months (one district every two months).

The routine work involves vaccinations against yellow fever, smallpox (which is no longer a problem), tuberculosis, and measles (for children between 6 months and 6 years old). The more vital work consists of discovering new lepers and in checking the progress of the disease in known lepers who are brought to us by the disulone distributor for the district. It also consists of eliminating sleeping sickness (trypanosomiasis), which has reappeared in post-independence Africa. We are using the most sensitive methods for biological research into trypanosomes, because the triple-centrifuge method is now outdated. In the zones where the ancient centers were (Comba), we have used the immunological method employed at Brazzaville. We also treated victims of intestinal parasites detected by our team of laboratory workers.

I believe we must increase our efforts to help those who suffer from all these tragic diseases, for the simple reason that they are human beings like ourselves—we who by the grace of God still retain our health.

Sonora Semiannulata, *vermilion lined ground snake*.

DOCUMENTING 181 AMPHIBIAN AND REPTILE SPECIES

Not many of us have enough tenacity and interest to embark on a project, work on it for 25 years, still consider it only 60 percent complete, and continue to make efforts to finish the job. In California, a remarkable man has been doing just this; he has been assembling what is likely to be one of the most unusually complete fauna studies ever done.

The pressure on the work is increasing in direct relationship with the growing urbanization of the Pacific Coast of the United States. With huge areas of the land giving way to bulldozers and construction crews, with new roads cutting erstwhile animal territories into ecologically unbalanced sections, a relatively neglected population of animals, the amphibians and the reptiles, already includes species officially cited as endangered. Scant public knowledge of and sympathy for these animals have made it difficult to gain support for their evident plight. Living away from population centers, not readily recognized, rarely appreciated, they are just as much endangered as many other, more publicly supported species. Bringing them and their situation into clearer public focus is the objective of this project.

NATHAN W. COHEN
1324 Devonshire Court
El Cerrito, California 94530
United States of America
Born October 3, 1919. American.

Two quite different positions occupy a major portion of the working hours available to Nathan W. Cohen. As director of the science curriculum at the University of California Extension at Berkeley, he develops and coordinates science classes for professionals in scientific and technical work—for example, medical technologists and radiation physicists. He also develops scientific symposia at highly technical levels for clinicians and research personnel in the biomedical fields.

His second position is that of research associate in herpetology at the Museum of Vertebrate Zoology of the University of California at Berkeley, where he participates in herpetological and ecological field research with colleagues in the museum and photographs field conditions and amphibians and reptiles for teaching purposes.

After receiving his A.B. (zoology) from UCLA in 1944, he earned his M.A. (zoology) in 1950 from the University of California at Berkeley, and in 1955 went on to take his Ph.D. (zoology) from Oregon State University at Corvallis. An expert photographer (photographic editor for the American Institute of Biological Sciences) and field scientist, he was the coordinator of the University of California at Berkeley "Galápagos International Scientific Project," managing the field activities of 60 scientists from nine nations on this six-week expedition.

I plan a documentary volume of color photographs and text about the most endangered species of animals inhabiting the Pacific Coast states (California, Oregon, and Washington)—the amphibians and reptiles. Many of these animals, vital links in our total environment, face extinction at our unwitting hands. The material will be aimed specifically at the general public and at educational and governmental authorities.

Project Description

My plans are to complete color photography of all the different species of amphibians (salamanders, frogs, and toads) and reptiles (lizards, snakes, and turtles) of the Pacific Coast states. I have photographed 60 percent of the approximately 181 species in these states during the past 25 years.

    The photography thus far accomplished has been done in my spare time. I hope that a grant will allow me to complete the photography of the remaining 40 percent. Particular attention will be given to those species classified by the US Fish and Wildlife Service as "rare" and "endangered," such as the black toad, black legless lizard, and San Francisco garter snake, as well as those species that are outstanding candidates to be so classified. The latter include the California desert tortoise, tiger salamander, arroyo toad, Panamint lizard, and mountain kingsnake. In addition to completing the species photographs, I plan to photograph the natural habitats of these animals whenever possible. The lack of unrestricted time in the field has often prevented me from photographing animals in their natural habitats, and I have had to faithfully reconstruct natural habitats in my home.

    In the field, I will focus on types of adaptive animal coloration, such as obliterative, disruptive, cryptic, advertisement, and warning coloration, as well as on other aspects of adaptation to the environment. Most herpetological studies have been devoted to taxonomy, morphology, and physiology, but much remains to be studied in the animals' adaptation to their particular niches in nature. The few studies of coloration that have been undertaken have yielded significant new knowledge about adaptation and speciation. My photographs will accurately document the animals in their natural habitats. Such recordings are particularly important in herpetological speciation research, which depends heavily on coloration and pattern. This is especially true for amphibians, which lose all or most of these characteristics when kept in preservatives. For example, the research on the evolution and relationships of the various groupings of the salamander genus *Ensatina* could never have been accomplished if the study had depended solely on preserved animals.

    Life histories—breeding, nesting, development of young, foraging for food, and other activities—will, where possible, be photographed and a narrative recorded. Although the photographic record and field notes will be scholarly and valuable to researchers, particularly if some of these species unfortunately become extinct, this is not a research project in an experimental or narrow sense. This project will result in a well-illustrated book written specifically for the general public and for students in high schools and institutions of higher learning. In addition to my own efforts and expertise, the vast resources of the University of California and of other repositories of knowledge are available to me. I also know the habitats and the herpetofauna of the Pacific states extremely well, so my time in the field will be efficiently used.

The major emphasis in both photography and narration will be an empathetic and esthetic treatment of the animals, their habitats, and their role in our total environment. Given my previous research, my experience in photographing amphibians and reptiles, and my field work, I am confident that I can complete all the photographic and narrative aspects of this project in one year of concentrated, uninterrupted work.

Rationale for This Work

There are many excellent college textbooks, field guides, and other herpetological publications, but none, to my knowledge, develops an empathetic and esthetic perspective on these animals, and none develops a sense of the doomsday that so many of these species face. Nor do any available publications relate with any consistency, if at all, an animal's habitat to its coloration, pattern, behavior, and other aspects of its existence. No herpetological book really stresses that the environment in which an animal lives is absolutely essential to its existence, that animals are inextricably tied to that part of the earth on which they naturally live. These are the environmental aspects I intend to document. The Pacific Coast states have a varied herpetological fauna, the richest in the western United States, but because of the ever-increasing threat to the existence of many species, there is an urgency to completing this project. The threat derives from people's need for more land, more natural resources, and more recreation. The demand for more land is converting our natural resources into fabricated objects that utterly destroy or alter natural habitats so that neither plant nor animal species can exist in these areas any more.

The most notable examples are large areas in the California deserts, now being decimated along with native plants and animals. This particular destruction is caused by recreational vehicles. Forest areas face similar fates when ecologically unsound timber-harvesting practices are employed. Even such a simple and essential activity as farming can threaten wildlife. Or the destruction of habitat may be closer to home: Housing tracts, parking lots, shopping centers, and highways now nearly cover a vast area south of San Francisco that was until just recently covered by ponds, grassland, shrubs, and trees. This area was the habitat of an abundant number of one of the most beautiful snakes in the United States, namely, the beneficial and harmless San Francisco garter snake (*Thamnophis sirtalis tetrataenia*). Now only a few of these garter snakes exist, near one spring, and this species is now on the verge of extinction. We destroy much herpetofauna out of ignorance, but wanton and deliberate killing of these species also occurs.

Even with such tremendous and seemingly hopeless odds against these animals, however, an educated and informed people can and will take steps to protect and preserve many of these unlikely candidates for public and financial support. This is evidenced by the fight for the Santa Cruz long-toed salamander. The general public in Santa Cruz, California, and elsewhere, along with schoolchildren, university students, faculty members, and state and federal agencies, after being informed about the impending extinction of this amphibian species, pooled their financial and political efforts and succeeded in protecting one of two ponds crucial for the survival of this species, and in purchasing the other pond from private owners at considerable cost. My photograph of this small salamander, which appeared in several local and national newspapers when the controversy over the ponds erupted between conservationists and developers, had a positive effect in molding public opinion for the preservation of this salamander. There are, of course, other examples in which an alerted public took a strong conservationist stand against private, state, or federal opposition and won the case; moreover, private business and others have sometimes modified or withdrawn their plans to accommodate various species of plants and animals. Thus there is clearly some realistic hope for the survival of amphibian and reptile species now on the brink of extinction.

Why is it necessary to educate and alert the public regarding the dangers these animals face? One reason is that lizards, snakes, and salamanders are mostly unknown to the public, and people in general regard them, when they regard them at all, as repulsive or as creatures to be feared. In addition, with some exceptions, the public does not generally know of the existence of these animals, even though many of them exist close to human habitation. Most herpetofauna, except for the giant tortoises and snakes one sees in zoos, are small and inconspicuous in nature. They are not spectacularly large and imposing like the threatened Bengal tiger nor as winsome as the helpless baby harp seal, which is brutally slaughtered for its pelt (a fact well publicized in the news media). While these and other well-advertised threatened resources should be protected, it makes little sense to care about exotic species, often thousands of miles away from our shores, and yet to pay no heed to our own endangered species—literally underfoot. Such lack of concern can only be ascribed to ignorance about the unseen wealth we have in our own forests, streams, grasslands, mountains, and deserts.

Why is it important to conserve our herpetofauna? They are part of our total environment, and, morally, they have the same right as we to survive. The beneficial roles they play for humans are often obscure. Yet, aside from the obvious benefits to people (snakes control grain-destroying rodents, and lizards prey on disease-carrying insects), amphibians and reptiles play a significant role in the food chain, feeding on smaller organisms and in turn serving as prey for larger animals.

Herpetofauna have also been important in research basic to human biology. Our understanding of our own being—reproductive processes, embryological development, the healing process of wounds, the nature of nerve impulses, the role of hormones, regeneration, and a host of other knowledge beneficial to humanity—has been mostly obtained from, of all things, frogs and salamanders! The venom of poisonous snakes is used to help us understand and alleviate pain. Studies on the mysterious pineal gland in our brains have been aided by studies on the "third eye" in lizards. Improvement in our night vision has been aided by research on nocturnal snakes. We do not know what further beneficial knowledge can still be gained by studying these animals—which is perhaps the most anthropocentric reason for conserving these species.

Nor do we yet know that all the species of herpetofauna in the Pacific Coast states have been discovered. My interest in applying for a Rolex Award is to help preserve these herpetological species so this and future generations can appreciate more fully their beauty, discover basic knowledge through research on them, and more nearly maintain the ecological balance of nature from which all living things on earth benefit. To these ends, the documentation of these species in color and narration will help to educate our society so the public will come to value and conserve the amphibians and reptiles of the Pacific Coast states as a public trust.

An amphibian or reptile being legally designated "rare" or "endangered" offers slight protection, if any, from individuals unaware or unconcerned of such status and bent on taking or killing such animals. This unfortunate situation exists because it is virtually impossible for law enforcement personnel to police all the areas these species inhabit. The only hope for the conservation of these species is an educated public. An informed public is the best protection these animals can have.

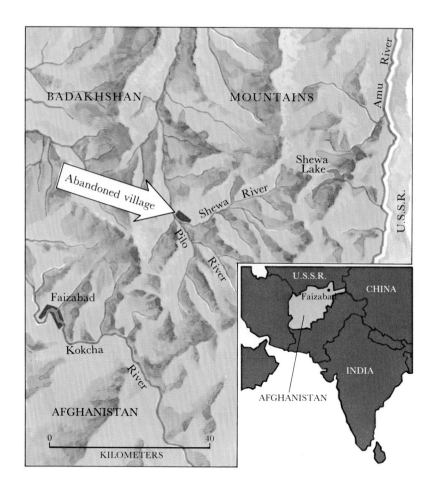

THE VANISHED VILLAGERS OF AFGHANISTAN — A TWENTIETH-CENTURY ENIGMA

For millennia, the lure of mystery and the unknown has led enterprising souls on travels around the world, often simply motivated by the desire to know "what happened." Although many of the great questions descend to us from myth and from times for which the records are lost, we do not lack modern puzzles.

Some of us today were alive when one of these modern puzzles arose, and so far no one has cracked the riddle. Entire villages do not normally and suddenly decamp for parts unknown; a few individuals may disappear from sight under decidedly peculiar circumstances—but 8000? All at the same time?

HIROSHI FUJII
Honorable Mention, Rolex Awards for Enterprise
20-3 Yazako-Kitsunebora
Nagakute-Cho, Aichi-Gun
Aichi Prefecture
Japan 480-11
Born March 11, 1940. Japanese.

Following his high school education, Hiroshi Fujii earned his B.Sc. in engineering from Shizuoka University in 1962. He then completed his M.Sc. (1964) and his Ph.D. (1968) at Nagoya University, both in engineering. At the time of his application, he was assistant professor of engineering at Gifu University, researching and lecturing in production engineering. One of his specialties is the history of technology (especially of traditional tool making) and of its propagation through Central Asia. This interest led him to the puzzling riddle that is the subject of his project.

I hope to discover why 8000 inhabitants mysteriously disappeared from a town in the Badakhshan Mountains in Afghanistan. At the junction of the Shewa and Pilo Rivers sits an abandoned village, of some 2000 houses, whose name even has been forgotten. It is known that the 8000 or so inhabitants all disappeared abruptly about 70 years ago, and it is said that their descendants are living in the border area between Afghanistan and Pakistan, although they sometimes rove temporarily. Nobody yet knows why the people left their original village.

In 1974, while on my way from Faizabad to Darwaz, I looked over at this village from a mountain ridge and saw a great number of abandoned houses. Ever since, the strange sight has made me wonder why the grassy glens surrounding the village forbid people to live there, even though the vegetation is rich only part of the year and only horse and donkey transportation is available. No one in this area seems to know the reason. I could find no literature on this mystery except confirmation of that disappearance in the *Historical and Political Gazette of Afghanistan* ("Badakhshan Province and Northeastern Afghanistan," 1972, vol. 1, published in Austria).

I and a number of people who are interested in this problem, however, have guessed a couple of possible reasons. Now, as a result of many fruitful discussions among these professors and specialists from varied fields, we wish to research the question thoroughly in Afghanistan itself. One medical doctor thinks that the inhabitants might have been suffering from a certain fever, a veterinarian and a zoologist think the domestic animals were affected by diseases, and two botanists think that grasses for animals or crops were destroyed by a violent change in climate. Furthermore, this area was one of the few regions where Shiah, an older sect of Islam that is now a minority, was firmly entrenched. If the surrounding tribes, who are Sunni, encroached on them, the villagers might have felt it necessary to abandon their homes in order to maintain the isolation their religion requires. Two members of our team, in the fields of genetics and linguistics, thus eagerly wish to find where the villagers' descendants live. Specialists using the special techniques of their disciplines hope to resolve the mystery. The close cooperation in research by various active and experienced experts will be one of the special characteristics of this exploration.

In order to manage this kind of project successfully, we need a leader who has had much travel experience in arid lands and who knows the customs of the native people. I hope that my experiences during four trips to Afghanistan will help our group adapt to the wild environment and to make use of all possible avenues of investigation in a very different culture.

I would like to make one more comment about the character of this project. Besides answering the basic question about the villagers' disappearance, the knowledge we gain through this project should also deepen our understanding of the conditions for human survival in a critical climate, for human history may be seen as a process of expanding habitation areas farther and farther into critical environments by means of developing improved techniques of tool making, agriculture, animal husbandry, and so on. We believe that our research will prove relevant to this broader concern as well as to answering our original question. We plan to collect the necessary data, analyze it, attempt to arrive at conclusions in both the particular and general areas discussed, and then publish our results.

HUMAN COMMUNICATION WITH GORILLAS

How far can we extend our ability to communicate with animals? There are no definite answers to that question now, but an ongoing study at Stanford University in California is pushing back the boundaries of communication in a fascinating way.

Because gorillas are unable to speak as humans do, their linguistic abilities can be studied through the sign language used by the deaf. Years of patient instruction have brought Koko, a female lowland gorilla, to the point of having a vocabulary that enables her to carry on a conversation at the level of a very young human child. She uses "words" and concepts in interchanges with her human mentors, in her own unaccompanied play, and with a new fellow student, Michael, a young male lowland gorilla. The implications of the study, given the possibility that the two gorillas may pass on the sign language to hoped-for offspring, are exciting in both a scientific and an educational sense.

FRANCINE GRACE PENELOPE PATTERSON
Rolex Laureate, Rolex Awards for Enterprise
700 Middle Avenue
Menlo Park, California 94025
United States of America
Born February 13, 1947. American.

As a doctoral candidate in developmental psychology at Stanford University, Stanford, California, Francine "Penny" Patterson has been pursuing a long-held interest in various aspects of the learning process.

An Edmund J. James Scholar at the University of Illinois (1965–69), she was elected to Phi Beta Kappa and Phi Kappa Phi in 1969 and went on to take her A.B. in psychology, also at the University of Illinois, in 1970. Her teaching and research experience in the fields of behavior genetics, cognitive development in pre-school children, adolescent psychology, and developmental psychology has led to the work in which she is currently engaged, as described in her project.

During a program initiated over four years ago, I established two-way communication between humans and the gorilla Koko at a level perhaps in advance of other studies of anthropoid apes learning a form of human language. And now I propose research with a second gorilla, Michael, a $3\frac{1}{2}$-year-old male; he will be a fellow student, companion, and future mate for Koko. This arrangement should considerably increase the scientific value of the project by confirming results with Koko. I will also have the opportunity to analyze individual differences in cognitive and linguistic development, intraspecific linguistic communication, and possibly the instruction and cultural transmission of gestural language in gorillas.

Project Koko is the first and only ongoing longitudinal study of the linguistic and behavioral development of a gorilla. It has now been in progress for almost five years (longer than any other study of an individual language-using ape). This research has multidisciplinary relevance; it transcends the fields of primatology, anthropology, psychology, biology, linguistics, and philosophy. I am striving to obtain comparative developmental records similar in nature and scope to those being compiled on human children (both deaf and hearing) and chimpanzees and to apply the linguistic and other behavioral data to elucidate the role of cognition in language development and use. Not only is this study contributing important information about the intellectual and linguistic abilities of this unstudied primate species, but it should also shed light on the origin and evolution of human language.

Summary of Progress

During the first 48 months of training, Koko, a female lowland gorilla, acquired a vocabulary of approximately 250 words in American Sign Language (the language used by the deaf in North America), which she spontaneously combines into meaningful and often novel sentences of up to 12 words in length. For example, Koko commonly makes statements such as "You chase tickle gorilla" and "Time you quiet sleep." At the age of 52 months, Koko frequently used up to 180 different signs during the course of a day. She is using this rapidly expanding vocabulary of signs in combinations that express semantic and possibly grammatical relations similar to those expressed by human children. These and other data indicate that the gorilla is performing at a level comparable, if not superior, to the chimpanzee subjects that are learning language. A unique finding has been that Koko comprehends novel statements in sign language and spoken English with equal facility.

The level of Koko's cognitive and linguistic sophistication has continued to advance. She spontaneously comments about her environment, feelings, and desires; answers "who," "where," and even "why" questions; invents words; talks to herself and uses language in imaginative play; has been known to lie; has a sense of humor; and has shown glimmerings of an ability for representational art, a talent thought to be unique to humans.

I am also finding evidence that the cognitive abilities of the gorilla have been underestimated by previous investigators. H. Knobloch and B. Pasamanick found little evidence of adaptive behavior in handreared gorillas beyond the 18-month level of human performance. The results of four different standardized intelligence tests on Koko during the past four years present quite a different picture. Her intelligence is now equivalent to that of a $4\frac{1}{2}$-year-old human, and her intelligence quotient (IQ) has remained constant—at about 85—over time and tests, much as a child's does.

Method

The new subject is a male lowland gorilla between three and four years of age. Michael, as he is called, shares a 10′ × 50′ house trailer with Koko on the Stanford campus. Both gorillas receive intensive individual attention from signing companions and teachers for 10 to 12 hours a day, seven days a week. Koko and Michael are gradually being introduced to one another, and when this process is complete they will spend several hours each day in a social situation with each other. This will allow me to document and analyze communication between gorillas, both linguistic and nonlinguistic. Because of differences in size, sophistication, and skills, it is possible that Koko will teach Michael.

Studying the communication between gorillas by sign language is already underway. Both Koko and Michael have addressed sign communications to each other, and in many instances these have been followed by appropriate behavioral responses. "Chase," "Come tickle," and "Catch" are frequent requests, and on one occasion Koko signed "Me hit you" to Michael, and then proceeded to carry out her threat.

An exciting prospect for the future is the possibility that Koko and Michael will pass on sign language to their offspring. Encouraging indications of this potentiality have already come from observations of both gorillas. Michael has actually molded the hands of companions untrained in sign language into signs—instructing them much as a parent might a child. Koko has structured imaginary social situations with her gorilla dolls in which she signs to them. For example, one day while engaged in solitary play in her room, she placed two large plastic gorilla dolls before herself, signed "Bad bad" to one of them, then "Kiss" to the other. Next, she signed "Chase tickle" and hit the dolls together as if making them play. When the play session ended, she signed "Good gorilla, good good."

In this phase of the study, comprehensive samples of the gorillas' signed utterances will continue to be taken with careful attention to the order of signs in constructions. Specific methods of data collection include longitudinal daily diary records, samples of all linguistic interchange within specified time intervals on both audiotape and videotape, and daily inventories of individual vocabulary items and sign combinations. This methodology is structured to provide records of the nonlinguistic context surrounding signing, of the linguistic interaction of relationships between successive utterances, and of comparisons of the gorillas' signing with that of their teachers, human children, and chimpanzees. These records, in turn, allow for analyses, comparable to those performed on children's utterances, of the grammatical and semantic structure of the gorillas' sign combinations to be undertaken. Dr. Patrick Suppes has provided the project with the unique opportunity to use computer programs and techniques developed for the extensive analysis of child speech at Stanford's Institute for Mathematical Studies in the Social Sciences (IMSSS).

The program also includes formal testing of the gorillas' linguistic abilities and an investigation of their relationships to their cognitive development in other areas. The following tests will be used:

1. Double-blind tests requiring the gorillas to identify objects and to describe relationships between objects.

2. Comprehension tests in which the gorillas are required to answer a series of questions about pictures or filmstrips illustrating grammatical relationships.

3. Tests of comprehension in which the subjects manipulate objects to act out instructions given by an experimenter (for example, "Put the clown under the truck" or "Make the dog kiss the cat").

Techniques for instructing Michael will be similar to those employed with Koko. Molding, or putting the subject through the action required for a sign, and imitation are basic methods used in the early stages of teaching a sign. Later, prompting can be more subtle, as with a touch on the proper part of the body or a verbal cue. In the final stage of teaching a new vocabulary item, the word is elicited by a picture or question, such as "What's that?" or "What want?" As with Koko, simultaneous communication, or spoken English in conjunction with sign language, will be used. This technique maximizes linguistic input to the subjects and allows assessment of the gorillas' abilities to process spoken language and to translate from the auditory to the visual mode. In addition, the modes are easily separated for tests of comprehension. We will routinely take samples of linguistic interaction in which signs only, or speech only, are employed by the experimenters.

Although speech production is not within the natural capacities of apes (because of differences in the structure and function of their vocal tracts), it is a new channel of communication that soon may be opened to the gorillas. IMSSS has suggested plans for an auditory language computer console that would give the gorillas this capability and would keep an accurate and complete record of all utterances produced in this mode. The console would initially consist of an array of 64 buttons depicting either pictorial or symbolic representations of the most frequently used signs in Koko's vocabulary. The console would be linked to the institute's PDP-10 computer and to a voice synthesizer. When buttons on the keyboard are pressed, the console would generate the English word equivalent through a speaker in the room. If a period button is depressed at the end of a sequence, the entire English "sentence" would be spoken by the computer. With this console and with the method of simultaneous communication, claims such as the following can be evaluated: "Apparently it is the child's innate capacity for auditory analysis that distinguishes him or her from the ape."

Language now enables secondary examinations of gorillas' intellectual, emotional, and social development at a new level: standardized intelligence tests, structured question-and-answer sessions, and numerous tasks derived from research on child and primate cognitive development.

Adding Michael to this study should extend considerably its scientific value by establishing that Koko is not an exceptional gorilla in her intellectual and linguistic abilities. I will be especially interested in comparing and contrasting the linguistic and behavioral development of the two gorillas. One reason for this is that differences between Koko and human children, or between Koko and chimpanzees, may be due either to a species difference or to individual variation. Data on Michael will help us decide between these two alternatives. Species differences, if found, can give us insight into paths the evolution of intelligence can take and into the fundamental nature of language.

"One would expect that human language should directly reflect the characteristics of human intellectual capabilities, that language should be a direct 'mirror of the mind' in ways in which other systems of knowledge and belief cannot." What I hope to show in my work with the gorillas is that we are not alone in that mirror, that through this new reflection of the gorilla mind we may come to a fuller understanding of ourselves, our language, and our biological heritage.

GETTING PROTEIN TO THE THIRD WORLD VIA SELF-HELP

In underdeveloped and largely agricultural economies, one of the most frequently encountered problems is that of inadequate diet, especially that of widespread protein deficiencies. The lack of sufficient protein leads to a variety of illnesses and diseases that aggravate already difficult social situations.

Attempts by governments and other concerned organizations to provide animal proteins for typically rural populations are often complicated by the lack of distribution channels, the high cost of management and expertise required to operate a meat-producing system, and so on. Nevertheless, those concerned with the problem continue in their search for effective solutions.

One such solution may lie in the approach described in this project, which takes the point of view that self-help may be the most practical and enduring answer. In essence, the solution proposed here brings together scientific knowledge of animal husbandry and the realities of village cultures on a level where the two can be combined with felicitous results.

ARUNA FERNANDO
Honorable Mention, Rolex Awards for Enterprise
Steggles Breeding Unit
Wallalong, via Maitland
New South Wales 2321
Australia
Born October 8, 1940. Australian.

A Sri Lankan (Ceylonese) by birth, Aruna Fernando received his early education at Royal College, Colombo, Sri Lanka, before going on to take his diploma in animal husbandry at Queensland Agricultural College at Lawes, in Queensland, Australia. Then he joined Steggles Poultry, Ltd., one of the largest fully integrated poultry processors in Australia; they process approximately 18 million poultry annually. His career there, only briefly interrupted by a 14-month visit to his home in Sri Lanka, has led to his current position as liaison officer for the entire breeding operation; that is, he coordinates activities of the nucleus (or basic) breeding unit (run by himself); breeder, brooder, and rearing farms; hatchery operations: and field operations of genetics and laboratory personnel.

Thus he brings considerable technical and operational expertise to his proposed project, as well as a deep understanding of the needs and problems of the people he would help.

I plan to provide a cheap and efficient source of protein—eggs (and occasionally, where custom and religion permit, poultry meat)—to the grassroots level of the Third World, at minimal cost to governments or any other aid givers. I designed this project as a simple cottage industry for villagers, who must be educated and encouraged to ward off protein deficiency diseases. Villagers should be encouraged to buy chickens, with or without a subsidy. This will create a sense of ownership, as the actual husbandry of these animals will be the responsibility of individuals. This sense of self-interest, I feel, is vital for such a project. The success or failure of this whole project depends mainly on the enthusiasm, and to a lesser extent on the ability, of the people concerned.

I plan first to provide a small nucleus of chickens from a reputable poultry breeder in Australia (or any other developed country) for a few pilot projects, located in carefully selected villages in the Third World.

The pilot projects are essential to the whole concept. I plan to set up these pilot projects and follow them through until enough results are achieved to analyze and appraise the whole idea.

Breed of Poultry to Be Used

In this project, providing eggs for human consumption is a primary concern and providing poultry meat is secondary. Hence, instead of considering a dual-purpose breed, I will concentrate on proven egg producers with other traits suitable to the intended environment. Although hybrid chickens are more vigorous, hardy, and tough, I do not think hybrids will satisfy the main requirements for this type of cottage industry.

First, as no artificial incubation of eggs is planned, or possible, the "broodiness" in the breed of hen is important. Second, provision of feed for the poultry will be a great problem. Hence, it is imperative to choose a breed that is well known for its foraging skill and ability to feed itself without household scraps and other types of food and grain. Third, resistance to heat is another quality that must be considered carefully. And, fourth, the color of feathers selected is also important; the people in Sri Lankan villages, for example, are accustomed to the wild jungle fowl (*Gallus gallus*), a highly colored bird. If an all-white bird were provided, acceptance might not be total.

Given all these points, a breed such as the Brown Leghorn would probably be most acceptable and appropriate. They are good foragers, have great heat resistance, and an acceptable color. One negative factor, however, is that not all of the hens go broody.

Setting Up the Pilot Projects

This is a vital step for a correct and meaningful appraisal of the whole project. In my country of birth, Sri Lanka, I will set about selecting pilot projects as follows. First, I will approach the Minister of Local Government to obtain his permission and, more importantly, his blessing for the whole concept; then, with his aid, I will get help from a few members of Parliament from four or five regions with diverse climates, customs, and religions to set up the pilot projects. In each of these regions, I will set up projects with specific participants.

1. The headmaster or teacher of the village school can use this as a school project and thus involve the very young, who are open to new ideas.

2. A youth organization can take on the project as a cooperative scheme, thus helping the youth to achieve something in life that their parents have not done.

3. With the help of a member of Parliament and the village elders or head, I will select two or three above-average households to be used in the project and to set an example for the rest of the village.

4. Finally, I will organize a small competition in one village and give chickens to all who wish to take part and thus create a willingness to achieve results.

My Part in the Project

I plan to coordinate the whole initial operation in setting up the pilot projects, that is, managing the operations of chicken breeders, the air freight of day-old chicks, and the agencies handling the delivery of chicks to selected villagers.

The initial impetus in setting up the pilot projects is vital for this whole concept. Done correctly, with an eye kept on detail, it is bound to create a good impression not only among participants but also in surrounding areas. If funds permit, I will spend some time in the villages prior to the arrival of the day-old chicks, demonstrating ideal methods of husbandry. This example is bound to be retained by a few of the more observant onlookers. This phase is only possible if funds permit and if my employer grants me leave of absence.

If funds allow, I also hope to obtain leave to visit each pilot project twice a year, staying at each village a few days. I will make an appraisal after each visit. More importantly, however, I can provide momentum and encouragement. During these visits, I can also impart a rudimentary idea of disease control, mainly preventive rather than curative.

Major Problems I Foresee

Feed for the Poultry. As food provisions are hard to obtain even for people, it is unlikely that villagers will have much spare feed for the poultry. At harvest time, when sufficient feed is available, an educational program may have to be set in motion, to encourage storing as much secondary grain for the poultry as possible. And it is essential that the breed of poultry selected be a good forager.

Continuation of Generation. A difficult problem is ensuring continuous generation. Some religions, especially orthodox Buddhism, discourage their devotees from consuming fertile eggs, even when only the slightest suspicion of such a situation exists. Hence my project is planned for more liberal villages.

At the outset, if male chickens are provided at a ratio of 1:10 females, the continuation of generation is assured, provided some hens go broody. No sophisticated incubation system is envisaged or possible. (Such a system requires heavy capital costs, maintenance costs, spare parts, trained mechanics for repairs, and so on.) Instead, if natural incubation of fertile eggs is encouraged by selecting a breed of hen of which some go broody, sufficient chickens will be produced for each household.

Acceptance of the Project. Owing to religious and other long-standing beliefs, it will be hard to sell this idea of self-help in obtaining such a vital requirement as protein. This will hold especially true for the idea of retaining fertile eggs for the next generation of poultry. A small nucleus breeding flock, together with roosters, may need to be run in an area where it will be accepted. Then the orthodox villagers will only have to select a few broody hens from their own flock to sit on the fertile eggs supplied from the outside.

The Future

In regions where pilot schemes receive a favorable response and where villagers are willing to pursue a more comprehensive program, I can supply them with chickens. The pilot project should take approximately 12 to 18 months before meaningful results can be obtained.

BATTLE CHESS

Although historians may debate its origins and antiquity, modern chess aficionados agree that the quintessential game derives much of its tension from a set of age-old rules. Widespread, long-held knowledge of the game has resulted in countless opening strategies, mid-game tactics, and end-game maneuvers. Retaining the game's lore is one of the characteristic goals of the accomplished player, resulting often in fairly automatic opening moves and responses. These tend to be tedious for equally matched, highly capable players but intimidating for the overmatched beginner whose knowledge of the standard practices is minimal.

Purists may reject any tampering with the noble game, but many will be intellectually satisfied by the idea of a slight alteration. If, as this entrant's project suggests, the proposed change were to enliven the game for experts and broaden the appeal to potential new players, even purists should welcome the innovation.

VASUDEVAN MOHANLAL
A. A. C., Range "A"
Income Tax Offices
Race Course, Coimbatore 18
Tamil Nadu
India
Born March 2, 1941. Indian.

After taking a B.Com. (1960) and a B.A. in history (1964), both from Madras University, Vasudevan Mohanlal went on to a post as assistant commissioner of income tax in the region of Ahmedabad, India. As a chess-playing hobbyist, his interest in the game's development and propagation resulted in the project he outlines here.

The ancient game of chess, which originated in India, is played throughout the world today. The game has been intensely and exhaustively studied, and most of the moves have been reduced to standard openings and stereotyped play. The great Cuban chess player Paul Capablanca once suggested a more challenging type of chess game with more pieces on a bigger board. I have invented such a game, one that challenges the intellect of the players, makes new moves and strategy possible, and appeals to the "combat instinct" and the ingenuity of the players, rather than to their textbook knowledge of the game.

The game of Battle Chess is played on a chessboard of 100 black and white squares, instead of the usual 64 squares. The chessboard has 10 squares on each side and is placed with the white square on the right-hand side. It can be made of cardboard or plywood.

The playing pieces can be mass produced in plastics or carved from wood. The pieces in the game are:

1. The commander is represented by a figure of a commander and the letter *C*. The object of the game is to capture the opponent's commander; the player who does so wins the game. The commander moves one step in any direction at a time. This movement is similar to that of the king in chess.

2. The bomb is represented by a model of a bomb and the letter *B*. Just as nuclear bombs are the most powerful weapons today, the bomb is the most powerful piece on the board. It moves any number of squares

at a time in any direction—horizontally, vertically, or diagonally. It corresponds to the queen in chess.

3. The missile is represented by a model of a rocket or missile and the letter M. It moves diagonally, any number of squares at a time, backward or forward. This movement is similar to that of the bishop in chess. The missile on the black square moves along black squares only; the missile on the white square, along white squares only.

4. The ship is represented by a tiny model of a ship and the letter S. It moves any number of squares horizontally or vertically at a time, but *not* diagonally. Its movement corresponds to the rook in chess. It can also be used in "sheltering" the commander, similar to "castling" in chess.

5. The tank is represented by a small model of a tank and the letter T. It moves two squares in any direction at a time. This is an entirely new piece and has no equivalent in chess.

6. The plane is represented by the model of a plane and the letter P. Its movement is in the shape of an L—two squares in one direction and then one square in another direction, or one square in one direction and two squares in the other direction. This movement is similar to that of the knight in chess. Like the knight, the plane can also jump over another piece in its way.

7. The soldier (army) is represented by a figure of a soldier in typical battle stance and the letter A. It corresponds to the pawn in chess. Like the pawn, it moves one or two squares on the first move, and then one square only. It can move forward only, except to capture an opponent's piece on one of the two forward diagonal squares. On reaching the last square opposite, it can be exchanged for a bomb or any other piece except the commander.

All other rules in chess regarding castling, check, stalemate, draw, en-passante, and so forth, are also applicable to this game. These rules can be found in any book about chess.

The following points of novelty make this game an invention. First, the board consists of 100 squares, compared to 64 in chess. Second, the pieces symbolize modern warfare instead of medieval warfare. This aspect updates the game's symbolism. And, third, there is a new piece called the tank, which has no equivalent in chess. The new piece and the bigger board make it possible to develop new openings and fresh strategies in the game.

THE LIGHTWRITER

There are no more pleasing inventions than those which overcome a human disability and provide a more normal life for those who are restricted in some way. And when the development comes from someone who is faced with the problem himself, it is even more satisfying because it is a reminder of human indomitability and enterprise in the face of great odds.

The enterprise and the will of this project's author reflect that kind of spirit. He offers an implicit challenge to all of us who strive to overcome obstacles.

TOBY CHURCHILL
Honorable Mention, Rolex Awards for Enterprise
20 Panton Street
Cambridge CB2 1HP
England
Born June 29, 1947. British.

On completion of Perse Preparatory School and Perse School, both in Cambridge, Toby Churchill passed 7 O levels and 2 A levels, prior to going on to obtain a pass degree in mechanical engineering at the University of Bath in 1972. He is managing director of Toby Churchill Limited.

I suffer from difficulty in speaking along with muscular deficiencies. At home, I use an electric typewriter, but out of doors my means of communication was limited to a card with the alphabet printed on it; I spelled words out by pointing to the letters with my finger. This method makes heavy demands on the reader's patience, and I felt very isolated when away from home.

I decided I needed a portable electric typewriter, and with the help of three friends I designed and built one. We were immediately asked if they were available, so we set up a four-person company (Toby Churchill Limited) to manufacture and market them. The company was started in December 1973, and to date (March 1977) we have sold 35 machines.

The machine has been enthusiastically welcomed by therapists, social workers, and other professionals working in this field. We have also received many letters of praise from users whose Lightwriters have completely transformed their lives, easing family stresses and opening up new horizons.

There is a small but continuing demand for these machines. A couple of similar machines are on the market, but these are geared to the able-bodied user and have small keyboards unsuited to people with motor difficulties.

One of the disadvantages of having such a small turnover is that we are unable to advertise and usually unable to attend exhibitions of equipment for the disabled. So the majority of the world is unaware that we exist. Even when people do hear about us, they cannot see the machine unless they travel to our works.

It costs about 50 pounds sterling to attend an exhibition in the United Kingdom and about 1000 pounds to attend an exhibition in Europe or in the United States. It costs about 200 pounds to advertise in a trade journal.

If commended for the Rolex Awards for Enterprise, we will set up a special publicity fund that will be separate from our day-to-day trading account. We will use the publicity fund specifically to make our machine more widely known.

Eichornia crassipes, *the water hyacinth.*

TURNING WATER HYACINTHS INTO AN OPPORTUNITY CROP

An old business adage concerns the idea of turning problems into opportunities by viewing them from different perspectives. Sometimes that's tough to do, and few concerned people who have observed and fought the suffocating spread of the water hyacinth are inclined to look at it as anything other than an evil ecological monster. So far, however, attempts to stop the almost cancerlike growth of a plant destructive to water resources have not met with signal success.

Perhaps the previous approaches have missed a vital aspect of the problem. Too often those most concerned with the plant's blight become concerned only after disaster strikes close to home, when the water hyacinth invades local waters and chokes off other needed plant life, upsets fishing, and clogs previously open waterways. By that time, it is often too late, and the cost of the battle becomes prohibitive, creating despair and frustration.

In this project, a solution is proposed that has a considerable ring of hope in it. The problem has been looked at in a new way, and its solution may turn out to be elegant.

GODOFREDO G. MONSOD, JR.
258-C Tomas Morato Avenue
Quezon City, Metropolitan Manila
Philippines 3008
Born January 15, 1925. Filipino.

As a businessman–inventor, Godofredo G. Monsod, Jr. is the president of three separate Philippine companies and a dedicated pursuer of the spirit of enterprise in the Pacific. He has participated in workshops in the Philippines and Japan on invention development and intratechnology transfer, the manager's course at the Institute for Small-Scale Industries in the University of the Philippines, and a planning workshop on Philippines inventions development, under the auspices of the National Science Development Board in Tagaytay City.

My project calls for the full utilization of the water hyacinth as raw material for different products: poultry, livestock and fish feeds; fibers for pulp and paper; particle, acoustic, and insulation boards; textiles, foods, medicine, and deodorants. This aquatic plant was investigated to determine its valuable properties. These have been properly identified and now are being processed as a possible answer to the worldwide problem for no less than 50 countries directly affected by the suffocating speed of growth of this aquatic weed.

Control and management are basic approaches to use the plant effectively for economic purposes. The following are detailed descriptions of how water hyacinths can be controlled and managed.

First, growth, movement, and ecology of the plant are studied, and then the economic gathering of the plant is implemented. In developing countries, where the water hyacinth abounds and grows with unbelievable speed, harvesting by an average family of five is still a profit-making venture, for it does not require special tools. In highly industrialized countries, such as the United States, mechanized harvesting is feasible. The water hyacinth is collected from the shallow water and allowed to wither along the shore. After washing and rinsing, the leaves and roots are cut from the stems and are dried either under the sun during summer or in ovens during the nighttime or rainy season. The withered stems are defibered and dried by air. Below I present useful products and the methods of processing.

Poultry and Livestock Feeds

The animal feed ingredients are prepared by first gathering the water hyacinth plant, washing the roots and stems to remove all adhering impurities, drying them separately in any convenient drier (preferably a vacuum drier) at a temperature of about 120° C, grinding and screening them (preferably in a classifier vibration screen No. 1, 80–100 mesh) to produce water hyacinth meal (roots and leaves) as the major feed ingredients, and then mixing about 20–90 percent by weight water hyacinth meal with other feed ingredients, such as corn or flour meal. Some vitamins, such as vitamin A from water hyacinth concentrate, also may be added to this feed formula.

The leaves contain 18.7 percent protein, 2.07 percent calcium, 0.54 percent phosphorus, 3.2 percent fat, 17.1 percent fiber, 11.3 percent moisture, 36.6 percent carbohydrates (nitrogen-free extract, NFE). The roots contain 11.8 percent protein, 1.03 percent calcium, 0.67 percent phosphorus, 0.5 percent fat, 7.9 percent fiber, 11.2 percent moisture, and 41.6 percent carbohydrates (NFE). The economic viability of this process has been carefully analyzed and demonstrated.

Fibers

The withered, mature stems, which originally contained about 50 percent water, are defibered by machine, to extract the cellulose fiber. The fiber has a 16.08 percent tensile strength and an average alpha cellulose content of 84.80 percent. The fiber length is 1.65 millimeters and the width is 0.018 mm.

The process for recovering cellulose fibers from water hyacinth stems involves aerating newly harvested stems; separating lumped cellulose fibers from the pulpy binding material with a decorticating machine; steeping the lumped fibers in boiling, saturated sodium chloride solution for at least 20 minutes; washing the fibers thoroughly with fresh, soft water at ordinary temperatures; and, finally, air drying the clean fibers. The recovered cellulose material is in the form of loose, fine fibers with an average cellulose content of not less than 84 percent by weight. This process is commercially viable.

Paper Pulp

The process requires newly cut stems of water hyacinth or water lily. Mature stems are particularly desirable because their strong fibers are not readily damaged by chemicals in the process.

The mature stems are air dried at ordinary temperatures for not more than 72 hours, after which the stems begin to decompose and can be more easily separated from plant fluids and the parenchyma or other pulpy materials. There is no need to defiber the stems unless one wants to increase the percentage of paper pulp recovered. However, fine fibers mixed with the pulpy parenchyma cannot be recovered and converted into paper pulp. To avoid expensive defibering operations, therefore, one can process both fibers and parenchyma pulp together.

The partially dried materials are shredded or comminuted by conventional means. They are mixed with water and a pulping chemical in a digesting solution of sodium hydroxide. It is preferable to mix them in a closed digester or suitable vessel having the usual apparatus for regulating temperature and pressure. During extended operations, digesting solutions recovered from previous operations can be reused to avoid unnecessary waste of chemicals.

The fully digested materials, or pulped product, is discharged from the pressure vessel into a processing tank where the pulp material is separated from the solution, either by draining or by a convenient and inexpensive sieve procedure. The recovered chemical pulp is washed with fresh water and any undigested solids are removed. The resulting pulp can be converted directly into paper or paperboard. If a higher-quality product is desired, the pulp can be bleached before conversion.

The physical characteristics of paper sheets made from the stem fibers, excluding parenchyma material, were as follows: brightness, 72.5; burst, 47.5; tear, 36.0; fold, 1031.0, opacity, 77.0. Considering that the raw material is cheap, I think the process of manufacture is commercially feasible.

Particle Board

In defibering water hyacinth stems, a considerable volume of parenchyma cells is accumulated as waste. These wastes, if not recycled into something useful, will contribute to waste disposal problems and environmental pollution. To reduce production costs, I recycled the waste parenchyma into economically useful particle board.

Making multilayered particle board from stems requires collecting and cutting mature stems into flake and splinter particles, drying these particles to reduce the moisture content to 4–6 percent, mixing 30–100 percent of the particles with about 1–70 percent wood particles (based on predetermined densities), spraying the mixed particles with a glue mixture of about 6–10 percent by weight of the mixed particles, pouring the glued mixture in successive layers into stainless steel cauls glue mixture of about 6–10 percent by weight of the mixed particles, pouring the glued mixture in successive layers into stainless steel cauls with wooden frames or steel molds, pressing the multilayered mat in a hot press for about 7–10 minutes at a temperature of about 160–170°C and a pressure of about 280–300 psi, and curing the resulting multilayered particle board at ambient temperatures for at least one week. A particle board $\frac{5}{8}$-inch thick with a density of 639 kg/m^3 is produced.

Because the raw material is a recycled waste product, this project is economically viable.

Wine and Alcohol

The water residue saved from water hyacinth leaves and the water extract from parenchyma cells can be recycled to make wine and alcohol. The process comprises recycling the juice extract from stems and leaves after defibering the stems for pulp and paper manufacture and after removing rich chlorophyll, vitamin, and protein concentrates from the leaves. The combined extract from the leaves and stems is settled in a settling tank and filtered. The juice is heated to a temperature of 70–90°C (but kept below the boiling point) to kill harmful microorganisms. The extract is cooled at ordinary room temperature, and 3–5 grams of yeast (brewer's yeast) along with 100–300 grams of sugar per liter of the extract are added. It is allowed to ferment for two to three weeks; daily shaking prevents carbon dioxide formation. Then the fermented extract is filtered through a 200-mesh sieve until most of the suspended solids are removed. The resultant wine is ready for drinking. The wine's sweetness can be increased by adding sugar syrup. The fermented extract also can be distilled to recover all the alcohol, which in turn can be used in antiseptics and other medical preparations, and also in beverages.

The commercial viability of this process is still under study.

Nutrients

The fiber content of water hyacinth leaves (17.1 percent) is rather high, and therefore the leaves are not recommended for feeding starter chicks. To eliminate the fiber and to derive the most protein from mature leaves, a protein concentrate can be extracted from the leaves. This concentrate is useful for all kinds of feed rations. Furthermore, this protein would directly benefit millions of undernourished people in developing countries if it were proven an effective protein and vitamin A supplement.

The process for extracting valuable nutrients (vitamins A and B, complexes of B_1 and B_2, niacin, protein, and chlorophyll) from leaves of water hyacinths or water lilies is composed of the following steps: cutting leaves from stems, collecting and withering the leaves from 48 to 72 hours at room temperature to reduce water content, crushing the withered leaves by machine, feeding the crushed leaves into an expeller machine to extract nutrients, storing the juicy extract in a settling tank to further reduce its water content, collecting settled solids by discarding the liquid portion, air drying the solid portion containing the concentrate at a moderate temperature range, from 20–50°C for one to three days, and pulverizing the concentrate in a grinding machine to a particle size that passes through a 50- to 80-mesh standard sieve.

When utilized for human food, the protein extract with essential amino acids can be used fresh, or as paste, or dried. I have baked tasty hyacinth bread with corn or wheat flour. Rice flour flavored with shrimp or meat and reinforced with hyacinth protein makes palatable fritters. Noodles and soup reinforced with protein have also been made successfully. Without doubt, the product manufacture is commercially viable, and the cost is within reach of the poor.

Deodorant

A small percentage (10–15 percent) of hyacinth protein extract was mixed in chicken feed to test the effect of the concentrate during metabolism. After 24 hours, the chicken droppings turned charcoal gray, and the usual obnoxious odor had disappeared. Thus the extracted protein concentrate, supplemented with other built-in vitamins and minerals of the water hyacinth, is a potential deodorant. The chickens thus fed remain strong and healthy under abnormally bad sanitary conditions; the hens lay eggs with zero mortality rate.

My theory is that inasmuch as the chicken manure is deodorized by the process, the same would hold true for human waste. The water hyacinth can be eaten in either bread or soup form or taken as a capsule or tablet with the right amount of dosage per day (not less than 53 grams of protein and about 5000 International Units of vitamin A per adult). The water hyacinth thus offers the poor a balanced diet, and when ingested into the body, an effective deodorant for human waste. Thus, human waste can be commercialized as an NPK-rich organic fertilizer. The poultry industry, which is a major source of air pollution, will be benefited. This process will be a major breakthrough in the field of hygiene and sanitation.

Controlling the water hyacinth infestation of 50 countries seems hopeless. *Reader's Digest* has referred to the plant as a "most exotic nuisance," the *Natural Science* Magazine calls it the "beautiful blue devil," and the *Time-Life* edition of *Nature Science Annual* (1975) calls it "beautiful but dangerous." It has wreaked havoc in the United States since its first arrival, from Japan, in Louisiana as a Japanese entrant in a flower show. The US government has used every known scientific means to control it, in vain. Today, our ingenuity faces the challenge of how to check the destructive and unbelievably fast growth of this aquatic weed.

The initial achievements listed above have not been supported by any foundations or institutions. In fact, my work was ridiculed in the beginning.

Only lately, after two years of persistence, has this project achieved recognition. Since then, I have been honored with two Panday Pira Awards, presented by President Ferdinand E. Marcos of the Republic of the Philippines for the years 1975–76 and 1976–77, for my breakthrough in the economic utilization of the water hyacinth. In its desire to solve the problem, the US National Aeronautics and Space Administration (NASA), through its project leader, Dr. William C. Wolverton, communicated with me two years ago and sent a representative, James L. Otis, to the Philippines "to learn more about your plant-processing enterprise." I have also been invited to visit various NASA centers in the United States. Similar commendations have come from the UN Food and Agriculture Organization, the state of Louisiana, and the United Inventors and Scientists of America.

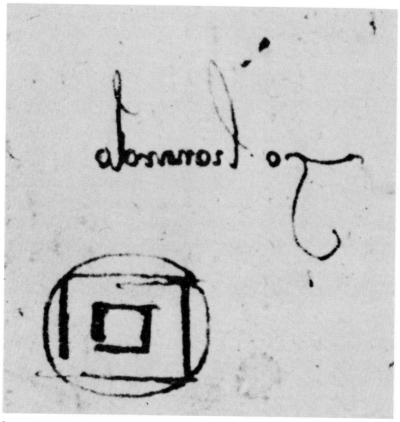

Leonardo da Vinci's signature, in mirror writing as was his custom.

THE SEARCH FOR LEONARDO DA VINCI'S "THE BATTLE OF ALGHIARI"

Through aged ledgers, diaries, letters, publications, and the like, art scholars can compile surprisingly complete logs of the works of early masters (whose masterpieces quite naturally tend to be preserved in any case). It is not difficult to imagine the professional consternation when a clearly identified "great work" appears to have been lost. When one compounds that consternation with trying to locate the work, the result is dismayed frustration. Such is the story of this project: the technological search for a major mural by Leonardo da Vinci. There seems little doubt that it exists, but the riddle of its recovery is almost worthy of Alfred Hitchcock.

JOHN FREDRICH ASMUS
8239 Sugarman Drive
La Jolla, California 92037
United States of America
Born January 20, 1937. American.

After taking his B.S. (1958) and his M.S. (1959) in electrical engineering, both from California Institute of Technology, in 1960 John F. Asmus went to the Technical University of Copenhagen to do ionospheric research. He subsequently returned to California Institute of Technology, where he took his Ph.D. in electrical engineering and physics in 1965. Following positions in industrial technology and in government, he assumed his present position as associate research physicist at the Institute of Pure and Applied Physical Sciences, University of California, San Diego.

He is a cofounder and member of the Center for Art/Science Studies, which was instituted to foster science in the service of the arts. Through the center, he directed programs to record deteriorating Italian sculpture via holography and to develop a laser cleaner for statues in Venice. He teaches laser theory, laser applications, and holography courses, and guides graduate students in research projects pertaining to lead isotope analyses of Benin bronze sculptures, transient heating for the conservation of leatherbound books, cleaning of paintings with lasers, electron-beam-assisted consolidation of friable stone monuments, and overpaint divestment with xenon flash lamps.

Leonardo da Vinci's great mural painting, "The Battle of Alghiari," has not been seen for at least 400 years. Historical investigations during the last few years have developed a strong case for the theory that Leonardo's masterpiece was not destroyed but was simply overpainted by Giorgio Vasari during the remodeling of the Hall of the Five Hundred in Florence, after 1563. With the approval of the Florentine government and the Italian Office of Culture, we have begun searching the strata beneath the enormous Vasari paintings for indications of Leonardo's work. The historical significance of the Vasari works dictates that nondestructive methods be employed in the search. After surveying and evaluating unintrusive advanced scientific methods (such as X rays, gamma rays, and microwaves), the search has now centered on adapting ultrasonic and infrared diagnostics to isolate the Leonardo painting and determine where to remove a portion of the Vasari mural to reveal "The Battle of Alghiari."

At the very heart of the Italian Renaissance is the Hall of the Five Hundred, in the Palazzo Vecchio of Florence. At the dawn of the sixteenth century, both Leonardo da Vinci and Michelangelo Buonarroti were commissioned to decorate this hall with the most magnificent mural paintings ever conceived. Tragically, by 1506 both Leonardo's "The Battle of Alghiari" and Michelangelo's "The Battle of Casina" were abandoned in an unfinished state. Technical difficulties, personal rivalries, political factors, and various disagreements all contributed to this outcome.

Nevertheless, the incomplete Leonardo painting, a group of violently clashing horses and men, was passionately admired in Florence over the following half century. Great painters, such as Raphael, had come to watch Leonardo at work; others, such as Rubens, copied his sketches or cartoons. This gathering of artists, which Benvenuto Cellini called "The School of the World," ultimately gave birth to baroque art, architecture, and music. "The Battle of Alghiari" disappeared from view by the middle of the sixteenth century. Its loss has been described by Cecil Gould, director of the National Gallery in London, as "a disaster comparable in magnitude to the blowing up of the Parthenon in 1687 or the Alcazar fire in 1734."

In succeeding centuries, it was assumed that what had briefly remained of Leonardo's work was merely a test panel or cartoon. However, during the past few years scholarly studies by my colleague Professor Carlo Pedretti and by others have proven that a much more substantial wall painting remained. First, Albertini's guidebook of 1510 records its presence on the wall in the hall. Second, a brief guide of the art treasures of Florence, published in 1549 by Anton Frencesco Doni, suggests ascending the stairs to the Great Hall and taking a diligent look at a group of horses and men "which will appear a marvellous thing to you." Third, the records of the city show that in 1513 a wooden frame was constructed to protect the wall painting, which, judging from the quantity of material, must have been roughly the size of "The Last Supper."

In 1563, the Florentine Grand Duke, unhappy over the unfinished state of the room, commissioned Giorgio Vasari to remodel the great, austere hall along its present sumptuous sixteenth-century lines. For 15 years, Vasari and his assistants worked on the project. No trace of

Leonardo's painting has been seen since then. The walls are now covered by Vasari's two cycles of Florentine military victories, each cycle measuring 8 meters by 44 meters.

For the past two years, my colleagues and I have been studying the structural nature of the walls around the Vasari paintings, as well as the floor of the hall. It now appears that Vasari had to modify the structure to make it a more regular shape, to increase the height of the ceiling, to make new windows and doors, to seal old windows and doors, and to accommodate the massive stone frames for his murals. Parts of the walls are lined with bricks; in others the original bricks and stones were hewn away.

Vasari said that Leonardo ran into technical difficulties, although that might have been Vasari's excuse for substituting his own paintings for the earlier one. But it is possible that the quality of the material available to Leonardo, who had invented a new technique, turned out to be inadequate for painting directly on plaster, the method he had attempted earlier with "The Last Supper." Since the latter had begun fading quickly, over a few short years, it is possible that by Vasari's time the "Battle" no longer remained sufficiently intact to make its preservation seem important.

It has generally been assumed that Vasari destroyed Leonardo's painting in order to execute his own "Battle of Val di Chiana." A crucial insight has been forthcoming from recent art conservation activities, which discovered that several Vasari mural paintings cover other artistic works. These include a Giotto fresco as well as Masaccio's "Trinity" in Santa Maria Novella. Vasari was quite careful to avoid damaging these works. He either built a false wall in front of them or covered them with lime before executing his own composition. Thus, it seems entirely possible that he might have done the same with Leonardo's painting.

Thus we are faced with the question of whether the Leonardo work is indeed preserved somewhere beneath the Vasari works, and if so, which 30-square-meter region under the 700 square meters of visible composition contains the remains of Leonardo's labors. For political, artistic, economic, and practical reasons, the search for "The Battle of Alghiari" must proceed without disturbing the present paintings. Consequently, for the past year I have been experimenting with nondestructive means of exploring the wall of the Hall of the Five Hundred. Laboratory simulations have considered X ray, gamma ray, acoustic, electric, and neutron activation analyses. We now believe that ultrasonic sounding and thermal imaging with infrared have the best chance of detecting any anomalies beneath the Vasari paintings.

During October and November 1976, the ultrasonic and infrared equipment was moved into the hall for preliminary tests. The ultrasonic equipment was found capable of detecting individual bricks, mortar joints, and detachments beneath the paintings. Infrared imaging tests have revealed bricked-in windows and doors. In December 1976, we constructed a small computer for recording the search data (ultrasonic and infrared) and analyzing the results. Some funding has been made available, and in April 1977 we shall begin our centimeter-by-centimeter exploration of the strata beneath the wall surfaces.

By the fall of 1977, we hope to be able to advise the Florentine government as to the prognosis for removing a portion of the Vasari work to uncover the "Battle."

The preliminary work in the Hall has already produced one spectacular discovery. We have found that only two words appear in the entire 700 square meters of Vasari's paintings. On a very small, obscure army banner are the words *Cerca Trova*, which loosely translates to "He who seeks shall find."

The green broadbill, found from India to Malaysia.

OPERATION JUNGLE SOUNDS

Those who never set foot in the exotic jungles of Southeast Asia miss one of the great nature experiences. Silent films can bring us the visual sensation of lushness and life, but much of the jungle can be imparted only through sound.

Thanks to this enterprising naturalist, we can listen to the sounds and voices of the jungle that we might never be lucky enough to hear. Armed with sophisticated recording equipment and a knowledge of his quarry's habits and habitats, he hunts the rare and seldom seen (let alone heard) birds and animals of a faraway world. He allows us to close our eyes and imagine ourselves where a mere film could never transport us, and the resulting closeness might lead us to protect the security of these endangered fellow animals, whose environment we are rapidly destroying.

GUSTAAF HERMAN TEEUWEN
P.O. Box 37198
Overport, Durban
Republic of South Africa
Born February 14, 1933. Dutch.

Despite being the proprietor of two retail shoe stores, Gustaaf H. Teeuwen has been able to find the time to pursue his long-held interest in recording the sounds of nature. Selections of his recordings of exotic birds, made in the jungles of Malaysia, have been played by the Swedish broadcasting organization and the British Broadcasting Corporation.

I propose to make a zoological expedition to Sarawak, Sabah, Kalimantan, (Borneo), and Sulawesi (Celebes) to obtain an original collection of high-quality tape recordings of rare birds and other animals found almost exclusively in the tropical rainforests of these areas. The undisturbed tropical rainforest of Southeast Asia, from the hot and humid lowlands to the mountain regions, contains a surprising range of flora and fauna. The forest is full of interesting and unusual sounds, from clear, melodious birdsong, interspersed with the calls of many more elusive babblers in the undergrowth, to the stentorian voices of large hornbills over the treetops, the exuberant "kuang kuang" of the rarely seen great argus pheasant in the depths of the forest, the spine-chilling shrieks and trills of broadbills, the weirdly melodic whisper of beautiful fruit pigeons, the monotonous "tonks" and trills of barbets during the hottest part of the day, and the bubbling waterfall song of the yellow-crowned bulbul near jungle streams. Raffles' malcoha, furtively moving about in the foliage, betrays its presence with a plaintive, catlike mew. The black-crested magpie, wandering in small flocks through the lower part of the jungle, boasts an unbelievable repertoire of melodious double gongbeats combined with goatlike bleats and thirst-provoking imitations of bottles running empty, while in the early morning hours the gibbons sing their unforgettable song. From the dark hollow of a fallen tree trunk, a large lizard, *Gecko stentor*, utters his ferocious grunt, which resembles that of a wild boar.

The forest avifauna of the Sunda Shelf, which includes the Malay Peninsula and the large islands of Java, Sumatra, and Borneo, shows a remarkable diversity of species, most of which are highly specialized for forest life. Because of this high degree of specialization, only a few of the rainforest birds proper have been found capable of survival in other habitats. The detrimental effect of mechanized logging on the flora and fauna in a complex ecosystem such as the tropical rainforest is considerable; it forces displacement of birds and other animals that cannot live and breed elsewhere, for even the largest forest is divided into territories, each with established residents who tolerate no intrusion.

Because Malaysia and Indonesia seem in a hurry to cut and burn their remaining natural forests, the aim of our project would be to save at least some part of the unique Southeast Asian avifauna for posterity. The birds themselves cannot be saved without the forest, as most of them would not live and breed in captivity. Therefore, we propose to tape as many as possible of their calls and songs before it is too late. These tape recordings can be made available to science and used for educational purposes in the form of discs or cassettes.

The territories being considered are Borneo (including Kalimantan) and Sulawesi (Celebes), both of which straddle the equator. Sudden tropical downpours and high temperatures are natural in this environment, as are leeches and a host of other pests, such as malarial mosquitos, large centipedes that can inflict a severe poisonous bite, scorpions, savage red ants, ticks, and biting midges.

Borneo, the third largest island in the world, is estimated to be two-thirds covered by primary forest. This estimate may have to be revised soon, however, for timber extraction by many international companies has recently made destructive inroads in Kalimantan. Borneo's avifauna, with a total of nearly 550 different species, is generally representative of the Malaysian bird fauna, but over 5 percent of the total is peculiar to Borneo. As far as we know, no ornithological sound recordings with professional equipment have ever been made in Borneo.

Sulawesi, separated from the Sunda Shelf by the Makassar Strait, appears to possess a breeding population of nearly 200 species of land birds. Most of these developed in isolation and are endemic to the island, which would make a first-ever recording expedition interesting.

The only recorder that could produce optimum results in this type of work is the Nagra, manufactured by Messrs. Kudelski S. A., Cheseaux-sur-Lausanne. The recorder to be used is the Nagra IVD, equipped with two special preamplifiers to accommodate an A.K.G. directional gun microphone. This microphone ensures full mobility in dense forest, and in my previous work I have used it to successfully replace the customary parabolic reflector. The recording speed is 15 ips on professional Agfa low-noise tape.

The Breogan, *August 1977.*

ACROSS THE SEA IN A SKIN BOAT

No one knows for sure just how contact was made in ancient times between the Celtic peoples of Britain and northwest Spain; only that crossing the seas in those days would seem unlikely. Nevertheless, the contact undoubtedly existed, and the riddle today rests on how it was accomplished.

Until now, our knowledge of early sailing craft and techniques of navigation in those ancient days suggests that such a sea voyage would have been nearly impossible. Today, however, it appears that we should be giving the early Europeans more credit for their courage, sagacity, and ability to build seaworthy boats from the most primitive materials. It seems they *could* have done it, and a dedicated group of nautical archaeologists has set out to prove how they might have. It is a voyage both across the open seas and across centuries of time.

FERNANDO ALONSO ROMERO
Honorable Mention, Rolex Awards for Enterprise
Departamento de Inglés
Facultad de Filología
Universidad de Santiago de Compostela
Spain
Born June 24, 1944. Spanish.

Fernando A. Romero took his bachelor of arts degree in philology (1970) and his Ph.D. in archaeology (1974) from the University of Salamanca. His present position is lecturer in the department of English at the University of Santiago. Fascinated with the world of ancient maritime travel, he is the author of the book entitled The Prehistoric Atlantic Relations Between Galicia and the British Isles, and Systems of Navigation. *His project pursues this long-held interest.*

The proposed project involves a sea crossing from Ireland to Galicia in a reconstructed Iron-Age boat of wicker and hides. The project is being undertaken by the team of experimental nautical archaeology at the University of Santiago (departments of prehistory and English). This team works on different aspects of the expedition. Groups of students are working on the history of prehistoric navigation and researching ancient methods of curing and tanning skins, archaeological proofs of prehistoric relations between northwest Spain (Galicia) and the British Isles, dietary customs of the ancient Celts, and Irish Celtic and Galician literature and legends.

The success of our project would constitute an enormous contribution to the history of navigation and of primitive contacts between Galicia and the British Isles and would form a launching point for studies of exploration and discovery. Yet the principal objective of this project is not so much the success of the voyage as academic research and experience to be gained by students in this field.

In the archaeological world, the primitive pre-Celtic and Celtic contacts between northwest Spain and the British Isles have always been a subject for discussion. Although it has long been known that they did take place, the exact nature of the boats that made these contacts possible has always been a mystery. We hope to solve this mystery by sailing from Galicia to Britain in a primitive boat made of wicker and hides.

After five years of research into the history of such boats in the Atlantic, we have constructed a faithful facsimile of the boats used by the ancient Celts (and even by the pre-Celts, considering its primitive structure). We have based our research mainly on archaeological finds, ancient systems of navigation, representations of boats in rock carvings (found in Galicia, Britain, and Scandinavia), information from classical (Roman and Greek) authors regarding skin boats, and dates supplied by Irish Celtic legends.

After having performed sailing tests on the *Breogan*, as our boat is named, we hope to depart from La Coruña in the second half of July, 1977, at the end of the academic year. We expect to make the crossing in less than ten days if the winds are in our favor.

We have received so far 50,000 pesetas from the University of Santiago and have been promised another 50,000 as soon as we launch the boat. We have also received 25,000 pesetas from the Nautical Exhibition at Ferrol and 10,000 pesetas from the Archaeological Museum of La Coruña. Recently, we have been given the promise of 35,000 pesetas from the County Council of La Coruña. We have made many requests for financial aid from various cultural bodies in Spain, but we have not received any more aid than that stated above. If we are unable to obtain further monetary aid, the team members will try to pay for the voyage out of their own pockets.

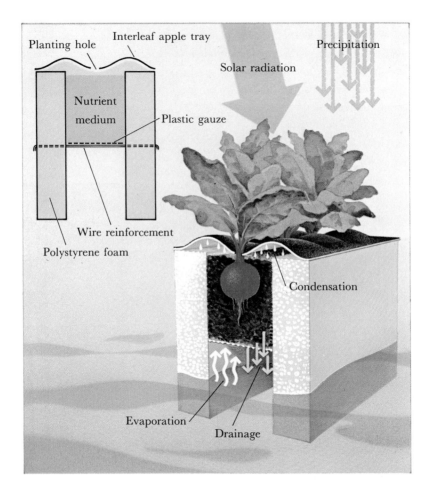

OCEANS OF BEETS

Trying to match an exploding world population with a dwindling supply of food-producing arable land is a problem that many thoughtful people perceive as an impending catastrophe. Until, or unless, we somehow can stabilize the world's population, simply providing a basic diet for all of humanity will continue to be dangerously difficult. Arable land is finite, and, while improving technologies can render it more productive, and political and economic realities may enable the conversion of marginal land to food production, we still cannot be sure we will be able to feed our grandchildren.

Because land appears to be a relatively inflexible factor in the food-production equation, perhaps we should resort to something in far greater supply on our harried globe—seas, inland lakes, and other large bodies of water. This project suggests a way to adapt our technology to water environments.

DAVID LINDSEY MILNE
Honorable Mention, Rolex Awards for Enterprise
84, van Wyk Street
Nelspruit 1200
Republic of South Africa
Born August 31, 1932. South African.

As entomologist and assistant director of the Citrus and Subtropical Fruit Research Institute in Nelspruit, Republic of South Africa, David L. Milne has some 73 scientific publications to his credit, continuing an interest that led to a B.Sc. in agriculture from the University of Natal (1954) and a Ph.D. in zoology from the University of London (1957). He has devoted professional attention to the population explosion and its effects on the environment. In this project, he describes a way to tackle one of the problems inherent in the future of that explosion.

I propose a new concept in crop production, using rafts that float on the sea. On these rafts could grow salt-tolerant vegetables, such as beetroot (*Beta vulgaris*), with a minimal need for freshwater irrigation. Prototype rafts have been constructed; on them beetroot plants can grow for up to 42 days without irrigation, because condensation and conservation of water are accomplished within the rafts. The potential use of this system for bays, lakes, and inland seas is discussed in the light of the current population explosion.

Background

Water covers 80 percent of the earth's surface. It therefore seems logical that the rapid growth in world population—at present estimated at 78 million per annum—should make us turn to the water surfaces of the earth as areas to be exploited for food production. As urbanization, industry, and transport daily consume large tracts of land surface, arable land is fast dwindling. Yet the water surfaces of the world are virtually unoccupied, except for the main shipping routes and fishing zones.

In addition, we are confronted with an ever-increasing shortage of water suitable for agriculture, despite the fact that we live on "the watery planet." During the next 20 years, municipal water consumption should increase tenfold, industrial water consumption fivefold, and water for energy production twentyfold. Yet little use seems to have been made of the simple physical laws of evaporation and condensation for crop production.

To solve some of these problems, a new concept has been developed, namely, floating vegetable beds, or "life rafts." Because of the salinity of the zone to be exploited, we grew common garden beets, *Beta vulgaris*, recognized as highly salt-tolerant.

Construction of Prototype Rafts

When this idea was first conceived, there appeared to be a worldwide accumulation of plastic waste that might be used to construct rafts. However, for our investigation we used expanded polystyrene because it is light, nearly waterproof, and easy to construct.

The basic unit of construction is shown in the figure. Plastic gauze as a base for the growing medium allows water vapor to rise through it and excess rainwater to drain through it. The growing medium itself was a mixture of half peatmoss, half kraal manure, with no additional nutrients. However, any moisture-holding, compost-type medium would be suitable, and supplementary feeding from above would also be possible.

The upper surface was covered with contoured plastic sheet, painted black, commonly used as padding between layers of packaged apples; the contoured surface design allows both precipitated and condensed water to accumulate in the zone of the plant roots.

The first prototype raft was placed in an outdoor tank of fresh water during March 1976, when temperatures ranged 20–30°C. Beetroot was sown on March 1, 1976, and the medium was irrigated with fresh water at this time. (Seed still germinates when 25 percent seawater is used, but speed of germination is retarded.) Germination occurred on March 8, 1976. For the first three weeks, only 8 mm of rain fell, yet the plants developed normally with no supplementary irrigation. Regular rainfall was recorded until the crop was removed in mid-April; the total rainfall recorded for the full growth period was only 25.7 mm. Seedlings kept in open pots indoors died of drought within one week after germination. It is therefore clear that the mulch effect of the surface cover, coupled with condensation from the water surface, greatly supplemented the water supply.

The second prototype raft, kept indoors at a temperature range of 14–20°C, was floated on seawater brought from the Natal Coast. These beetroot plants grew for 42 days without any supplementary irrigation before any signs of water stress were shown, while seedlings grown in open pots died of drought in ten days.

Conclusions

Table beetroot can grow well in 50 percent seawater and can even produce acceptable yields when 75 percent seawater is used. Thus, if condensation and precipitation are used as well as supplementary water sources, as shown in these experiments, we could attain an almost complete reliance on seawater as a water supply for food production.

Although the present study cannot be regarded as more than exploratory, it does indicate a potential area of development for crop production. Much of the world's rain falls on the sea, so we need not regard the seas as arid (without rainfall) areas. In addition, with this technology massive inland lakes all over the world could be exploited as cultivatable areas. Similarly, in coastal areas seawater could be directed into troughs or beds where rafts could be placed or could even be maintained above seawater-saturated sand. An element of self-maintenance could also be envisaged with beets, in which the harvested tops could be composted as a future growing medium; then part of the root crop could be used to produce alcohol, which in turn could power pumps for supplementary irrigation.

It is clear that a whole new technology would be required to fully exploit this concept and to deal with the problems of harvesting and wind, tide, and wave action. The population explosion, however, justifies development of such technology.

TRACKING THE GREAT WHALES

Although widespread international attention has focused on the threat humans pose to the existence of great whales, and encouraging international agreements have been reached in an effort to protect these huge creatures, their future is by no means assured. Disagreements between nations, whaling industries, and conservationists frequently center on disputed estimates of actual whale populations, with industry estimates ranging higher and conservationist ones ranging lower. Both groups, though, tend to share a mutual concern for obtaining accurate census figures.

Because the great beasts migrate over vast distances, obtaining reliable census data is virtually impossible. Now, however, ingenious imagination and advanced technology offer a method that could provide facts for those who need them. The first step involves plotting migratory routes. In this project, the whale-monitoring system would be linked to an observation satellite.

ROBERT M. GOODMAN
7811 Mill Road
Elkins Park, Pennsylvania 19117
United States of America
Born November 21, 1920. American.

As principal scientist in the Bio-Technology Laboratory at Franklin Institute Research Laboratories in Philadelphia, Robert Goodman applies physical science techniques to sensing and measurement problems in the life sciences. He works with biologists, medical doctors, physiologists, botanists, and others to solve problems, develop instrumentation and systems and direct their fabrication where necessary, evaluate equipment performance, and sometimes—as in this project—take part in field activities and other applications.

A graduate of the University of Pennsylvania, where he took his B.S.E.E. degree in 1943 from the Moore School of Electrical Engineering, Goodman is the author of 33 papers and scientific reports. He holds US patents in the fields of medical instruments, electro-optical devices, physiological electrodes and electrolytes, electromagnetic counters, and subminiature data-recording devices and systems. He has chaired and co-chaired scientific meetings concerned with wildlife monitoring and brings that experience and interest to the project he describes here.

The overall project, of which this proposal is an interim step, involves attaching a harmless instrument-bearing vest to members of the great whale species. These instruments digitally record dead-reckoning paths and biological, behavioral, and ecological data. The instrument pack records up to 15 bits of data roughly every four minutes for a full year. At the end of the year, the instrument-bearing vest is released automatically, leaving the animal unencumbered. The vest and pack then float to the surface for recovery. Along the migration track, a radio in the surfaced pack periodically transmits a signal to an orbiting satellite. Such beacon positions are later used as fixes to rectify dead-reckoning headings accumulated in the "on-board" recorder.

The satellite provides location data that enable a recovery team to home in on the pack. The recorded data are coded by computer, and biological, behavioral, and ecological assessments are made. Of final and paramount importance is the fact that definitive knowledge of migration paths by species and subpopulations will allow the community of nations to take a census. This system is now made technically feasible by the "location" satellite and our subminiature recorder.

The project involves designing and fabricating an improved expandable, releasable, instrument-bearing vest for gray whale calves; designing and applying an instrument package to be carried on the whale vest; capturing one or more gray whale calves for application of the instruments; releasing these instrumented animals in their natural habitat and tracking them for a period of 15 days; recovering the equipment as it automatically releases itself from the subject animals; reducing the 777,600 data bits recorded in the data pack by computer; and publishing the results.

To consider the proposed effort, one may view the undertaking as subdivided into several coordinated programs, each involving continuous, intimate liaison between the physical and life science investigators. These programs are described as follows:

Instrumentation Program

Swimming Speed Measurement. The sensor will be based on a prefocused, pulsed Doppler system or a crossed-field velocity sensor. Selection of the preferred approach will be made within the first two months of effort. The final device will be miniaturized and optimized with regard to power consumption.

Magnetic Heading Measurement. Requires a miniaturized, multigimbaled, damped, magnetic compass with optical pickoffs; suitable integration will be built into the circuitry to determine dead-reckoning heading.

Pitch Angle Measurement. Involves a miniaturized pendulum sensor. This datum is needed to find the whales' path in the horizontal plane.

Water Temperature. We have built this sensor and its circuitry and have used it successfully in the field; it will be integrated directly into the system.

Water Pressure (Diving Behavior). We have developed a device to measure diving behavior and used it in the field; it too will be built into the system.

Time into Experiment. This measurement will be based on quartz-controlled, digital clock sources already used by us in the past.

Miniature Data Recorder. This tiny, incremental magnetic tape recorder has a capacity of about 2.5×10^6 8-bit words; its average power consumption is about 15 W. It will be suitably modified with regard to tape transport system as determined by our 1973–74 field studies. We will also improve the tape-incrementing mechanism. All data of interest will be recorded on this device—129,600 bits per parameter, or 777,600 total for this study.

Control Circuitry. A straightforward extrapolation of previous work.

On-Board Electrical Power. We will use lithium fluoride cells because of their high-energy density and excellent operating temperature range.

Instrument Housing. It will be somewhat ovoid in shape for strength and to minimize water turbulence. It will probably be fabricated from a fiberglass and foam composite, such as a special Airex.

Tracking Transmitter. A transmitter made by OAR, Inc. (San Diego, California) originally developed for porpoise studies will be adapted for this application. It transmits only when its antenna is clear of the water and is essential for tracking and recovery of the data pack.

Automatic Release Mechanism. A subminiature, micropower, quartz-controlled timer will be designed to fire a harmless, miniature squib to release the instrument-bearing vest from the whale; time accuracy of 360 hours ± 5 minutes will be adequate.

Whale Vest. Although the instrumentation per se will be designed at the Franklin Institute Research Laboratories (FIRL), this vest (an improved version of the one used in our 1973–74 expedition) will be designed at the University of California at Santa Cruz (UCSC). Because of its nature—contact with the animal, the necessity to expand with animal growth (20–90 kilograms per day), and the necessity to contract as the animal dives under pressure—the vest is best designed by the cetologists who know most about the animal; thus Dr. Kenneth S. Norris' team will produce it in liaison with my team.

Fabrication of the System

Insofar as possible, all electronic components will be solid state and micropower. Great care will be taken to "ruggedize" all packaging and to use materials that can withstand the hostile environments involved. All equipment will be stringently tested before field use, and adequate safety measures in design are mandatory. Housings will be tested hydrostatically at pressures exceeding anticipated diving depths.

Field Experiment

Necessary approval permits for the capture of the gray whale calves will be requested from the US Department of Commerce and from the Mexican government. We anticipate no unusual problems in this regard. The site of our work will probably be Magdalena Bay, Baja California, Mexico, or Lopes Mateos, Boca de Soledad, which is part of the upper Magdalena. These areas are gray whale calving sites from December to February. Field work and staff will be directed by Dr. Norris, with the support of Dr. Bernardo Villa-Ramirez (our colleague from Mexico), and myself. A ship capable of tracking the instrumented animals at sea will be necessary. It must be on site for at least 15 days following the capture and instrumentation. Adequate provisions and camping equipment will be necessary for our combined FIRL-UCSC-Mexican team during the capture and instrumentation phase. A suitably fitted trawler, captained by an experienced marine mammal trapper will be chartered for the capture effort. During the tracking and recovery phases, it is likely that only the three investigators (Goodman, Norris, and Villa-Ramirez) and the ship's normal crew will be aboard.

Data Reduction and Publication

On retrieval of the data pack after its release from the whale, the unit will be turned off and returned to FIRL. Data will be transferred from the pack recorder to our computer facilities and reduced for evaluation by our cooperating teams.

We expect to be able to publish unique data and performance information in the fields of whale biology, behavior, technology, and systems technology. We believe these contributions will be of significant value to workers in many countries.

Planning

The successful completion of the proposed effort will permit planning for the final phase of research with the gray whale. This will require tracking many subjects over their entire annual migration. It will use a satellite that can rectify dead-reckoning data and locate instruments following their release. Because the gray whale is known to move past Canada toward the Bering Straits, we will invite investigators from Canada and the USSR to join with us in that phase of the effort.

We also hope to study the sperm, humpback, and right whales, as well as others. The instrumentation system for these animals will be essentially the same as for the gray whale.

The project is expected to start in the fall of 1977. All gear should be developed, fabricated, tested, and ready for sea by January 1979. Field work will be undertaken in January and February 1979. Data reduction, assessment, and reports will be completed by the summer of 1979.

TURNING WASTE HEAT DIRECTLY INTO ELECTRICAL ENERGY

New solutions to our energy problems are being proposed daily, many being concentrated on the amount of power that can be obtained from a given source. Naturally, if you are talking about the need to move an automobile or some other work function, relationships of power to size are important.

There are other applications, however, where the size of the power unit is of less concern than the efficiency of its production. In a society conscious of energy costs, the *efficient* production of power takes on considerable importance. If we could convert "wasted resources" into energy at little or no conversion cost, we would be tapping a remarkable barrel of opportunity. In this project, a highly qualified scientist poses just that prospect.

MING LUN YU
Building 480
Brookhaven National Laboratory
Upton, New York 11973
United States of America
Born August 21, 1945. Chinese.

After receiving his physics B.Sc. and M.Sc. in 1967 and 1969, respectively, from the University of Hong Kong, Ming Lun Yu took a second M.Sc. in physics in 1971 at California Institute of Technology, where he later went on to earn his Ph.D. in physics in 1973. He is an associate physicist in the Superconductivity Group at Brookhaven National Laboratory.

The object of this novel proposal is to construct a device to convert waste heat directly into electricity with an efficiency approaching the maximal theoretical (Carnot) efficiency. The problem with waste heat, as with any low-temperature heat source, is that most of the power spectrum lies in the low-frequency end of the blackbody spectrum. This range of the frequency spectrum, 1 Hertz to 10^{10} Hertz, can be utilized easily with simple electrical circuits.

To make this feasible, fluctuation theory is combined with semiconductor thin film technology and a special materials-processing technique. The end result will be a sheet of high-density electrical components (10^{13} per square meter). The device will be cheap to produce, and if the expected efficiencies are realized it will offer an attractive alternative to solar cells for the conversion of sunlight to electricity.

The starting point for this discussion is heat noise. Any resistor at any temperature develops a noise power. If two resistors are wired in parallel, the second law of thermodynamics tells us that nothing will happen. However, if we heat up one resistor while keeping the other at a lower temperature, power is transferred from the hot resistor to the cold one. The net power (P) transferred is given by the equation

$$P = 4k \, \Delta T \, \Delta F,$$

where ΔF = bandwidth, k = Boltzmann's constant, and ΔT is the temperature difference between the two resistors. This net power transfer produces an additional noise voltage (V) across the cold resistor (R_{cold}), expressed as

$$V = \sqrt{4k\,\Delta T\,\Delta F\,R},$$

for the simple case $R = R_{hot} = R_{cold}$. This voltage is completely *independent* of the size or geometry of the resistors! Consider the following values: $r = 10^{13}\Omega$, $\Delta F = 10^{10}$ Hertz, $\Delta T = 50°C$; then

$$P = 5.6 \times 10^{-11} \text{ watts}$$
$$V = 25 \text{ volts}$$
$$I = 2.25 \times 10^{-12} \text{ amps.}$$

This is not very much power. The characteristics, however, are independent of the resistor's size. The total power would be 600 watts if 10^{13} resistors were available. This voltage is fluctuating, and, if 1000 resistors were wired in series, the power would be exactly the same as if there were only one resistor—that is, 5.6×10^{-11} watts—because at each instant in time the voltage of half the resistors would cancel the voltage of the other half. If these resistors were wired in parallel, the situation would be exactly the same. Another circuit element, a diode, is needed to allow the device to perform useful work. A little more thought shows that the diode is quite fundamental to the circuit. Without the diode, the power delivered to the load (cold resistor) would not do any useful work. The load would simply heat up. The diode allows useful work to be done by converting the power spectrum from a random alternating current to a pulsating direct current voltage. The second law of thermodynamics states that the efficiency of any heat engine operating between two temperatures (°K) is given by

$$\text{efficiency} = \frac{T_{hot} - T_{cold}}{T_{hot}}.$$

The values $T_{hot} = 350°K$ (75°C) and $T_{cold} = 300°K$ (25°C) imply an efficiency of 14 percent. The power that does not appear across the load appears as heat across the diode. There is no such thing as a perfect diode. Any diode has a finite ohmic impedance at any temperature above absolute zero, and the best possible diode is limited by the second

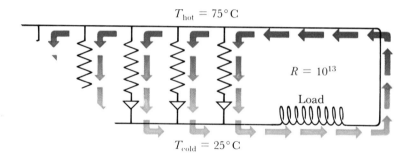

law of thermodynamics. The junction diodes now available are within 2 percent of the maximum theoretical efficiency.

The device's fundamental operation is now clear. It processes the low-frequency end of the electromagnetic spectrum emanating from the hot body (resistor). If a diode were attached to each of the resistors, and all the (10^{13}) diode–resistor components wired in parallel, a half-wave rectifier would result.

How can a practical, cheap device be built? There are any number of fabrication techniques, but only a simple one will be discussed here. A metal plate—say, aluminum—is covered with an appropriate eutectic mixture of quartz and silicon. (Whether this particular eutectic is now available is not important: Finding the right eutectic combination is an important phase of the investigation.) Removing the heat through the aluminum plate cools the mixture through the eutectic temperature and results in a quartz matrix containing an array of silicon rods. These silicon rods will be rather uniformly spaced and will lie perpendicular to the aluminum plate. The area of the rods can be as small as 10^{-10} cm². This process is known as *unidirectional solidification through the eutectic point*. The resistance of each rod can be estimated as follows:

$$R = \frac{1}{A} = \frac{10^3 \Omega \, \text{cm} \times 1 \, \text{cm}}{(10^{-5})^2 \, \text{cm}^2} = 10^{13} \Omega$$

assuming $\rho = 10^3 \Omega$ cm for the silicon and $l = 1$ cm is the thickness of the matrix coating, which is also the rod length. A diode must now be formed on the exposed ends of the silicon rods. To accomplish this, the whole assembly is put into a furnace and exposed to the proper metal vapor—say, indium—creating a thin (~ 100 Å) p-type layer. The p-type layers are next coated with a thin film of n-type germanium, creating a junction diode. A final coat of aluminum is evaporated over the matrix diode surface, forming the cold junction electrode, and the device is complete.

Analyzing the device again for $\Delta T = 50°C$, $R = 10^{13}\Omega$, and $\Delta F = 10^{10}$ Hertz (not an unreasonable figure for a microwave diode), we find

$$V = 25 \text{ volts}$$
$$I = 24 \text{ amps}$$

and

$$P = \tfrac{1}{2}VI = 300 \text{ watts/meter}^2$$

assuming 10^{13} diode–resistor combinations per square meter. The $\tfrac{1}{2}$ appears in the expression for power because the device is a half-wave rectifier, and only half the power can be delivered to the load. All of the preceding discussion has been phrased in terms of power. The efficiency of the device ultimately cannot exceed the Carnot efficiency (second law of thermodynamics), and the device should be analyzed for energy efficiency, not power. Drawing maximal power from a device generally decreases the efficiency, but to find the exact efficiency would require a detailed mathematical analysis, which is not necessary and would just clutter up the discussion.

Consider a solar energy application. The solar constant is approximately 1200 watts per square meter. As shown earlier, the maximal possible efficiency for the device with $T_{hot} = 350°K$ (75°C) and $T_{cold} = 300°K$ (room temperature) is only 14 percent. Therefore, under these conditions the device can deliver no more than 150 watts per square meter. Of course, if 100 of these devices covered a roof (100 square meters), substantial amounts of power would be available. Another alternative would be to use reflectors to concentrate the sunlight on a single panel, raising T_{hot} and thereby increasing the efficiency.

A second loss of efficiency in the device is caused by the finite thermal conductivity between the hot and cold plates. This loss can, however, be made as small as desired by making the device thicker. For example, ten plates could be stacked and wired in parallel. Assuming a thermal conductivity of 10^{-3} watts/cm-°C, and a temperature difference of 50°C, for a plate 1 square meter by 10 centimeters thick, the thermal loss is 50 watts.

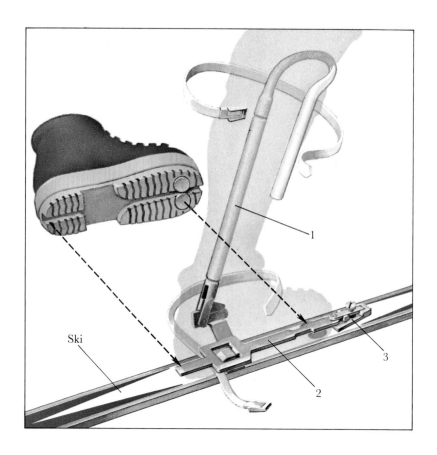

SKI BINDINGS WITHOUT BOOTS

Glorious and exhilarating sport that it is, skiing nevertheless exacts a certain tribute from its enthusiasts in the form of heavy, expensive, and uncomfortable boots. The design of conventional ski boots is based mainly on the requirements of modern ski bindings, which come in a variety of mechanical systems. Boot and binding together must perform the twin functions of providing control over the ski and a measure of safety in the event of falling—an engineering feat of considerable delicacy and precision.

Few skiers would argue that the ski boots of today are comfortable in all wearing conditions. Although they may be well suited to downhill running, the boots are not designed for ski walking or ordinary "off-ski" walking when heading for the slopes or taking a lunch break at the top of a mountain. Nor would most skiers swear by the "absolute safety" factor in their boot-and-binding combination. Although design has improved considerably over the last couple of decades, casts and crutches are still a common sight at all ski resorts.

Now, thanks to the enterprise shown in this project, much of that may change. Not only may the boot-and-binding combination become much more comfortable and safe, but the costs involved may tumble!

ANTONIO FAULIN
Honorable Mention, Rolex Awards for Enterprise
via Giovanni da Procida, No. 4
Milano 20149
Italy
Born January 3, 1933. Italian.

Having earned his doctorate in architecture in 1960 from the Politecnico di Milano, Antonio Faulin now practices as an architect.

The present invention is a new ski binding that negates the need for the usual rigid boots. This new device allows skiing with a "normal," comfortable shoe with a soft upper and an all-purpose sole. The sole no longer must be pressed down on the ski.

The device has three elements: (1) an elastic "arm" covered in soft material to be fixed to the leg at the shin bone; (2) a steel plate to be fixed to the foot, (3) a disengaging block, in rubber and steel, to be fixed on the ski (see the figure). All three items are interconnected to form a mechanical articulation that, together with the legs, absorbs all dangerous strain.

The construction is very simple, with few mechanical items, inexpensive parts, and great ease of use. It is possible to mount and demount this device on the ski without a tool, and because it is made of light material, the weight can be less than 1 kilogram (2.2 pounds).

In the eventuality of a fall, the device protects the leg from possible shock. All parameters affecting the release of the ski during falling are minimized; the release takes place between small steel surfaces at high pressure, rather than between big surfaces of different materials at low pressure, as in the usual system. Loosening in all directions, it is operated by only one "snap action" element, which has micrometric adjustments that unlock with a maximal breakaway effort depending on the skier's weight and ability.

It distributes moderate strains to anatomically suited parts and completely eliminates bending the shin bone, thus protecting it from shocks. All movements become more natural and easy, and with the same ski one can do alpine skiing. There is no danger of release during skiing. A few prototypes of this device have been built and tested by experts with good results.

Need for the Invention

At present, normal skis are connected to the skier's leg or foot with special bindings that grip the ski boots firmly, but release when tensile strength, torsional stresses, or any stresses become dangerous to the leg. The footwear has two functions: (1) binding the foot to the ski with a rigid sole, especially shaped and rounded; and (2) limiting movements of the leg, shin bone, and fibula with respect to the foot and consequently with respect to the ski, thereby partially blocking the ankle.

This latter function is undoubtedly a result of developments in skiing, particularly downhill and pleasure skiing equipment, but it also has been dictated for safety reasons. Because the ski footwear has the task of discharging muscular forces from the leg to the ski, it requires a certain shape, rigidity, and weight for the ski boot that causes considerable discomfort to the skier. In fact, the current shape, weight, and rigidity of ski boots constitute a serious inconvenience for the person who wears them not only while skiing but also for short walks to ski slopes, transport, and so on.

One objective of my invention is to create skiing footwear having uppers, and eventually soles, that are soft and flexible; in other words, footwear for skiing that maintains the characteristics of normal footwear for walking, thanks to simple expedients that do not in any manner alter comfort or practicality.

Another objective of my invention is to provide a ski binding wherein the use of soft shoes eliminates the need for a rigid center of the ski because of a rigid boot sole. This alteration increases the flexibility of the ski and remarkably improves technique, even in competitive skiing.

My invention provides a ski binding that allows the recovery of normal articulations; that is, it enables skiers to control the articulations of the foot and leg. It also helps them better and more effectively transmit muscular stresses from the leg to the ski.

A further objective is to create a ski binding that is easily and rapidly folded on the ski itself, so as to limit its bulkiness and to facilitate transport.

A final objective is to provide a ski binding that can replace the high, rigid binding–boot combination with a more effective and practical device, and above all with greater safety for the skier.

Description of Leg and Foot Positions

It is useful to examine the movements of the foot and leg on skis with reference to the equilibrium position, keeping in mind the positions assumed while skiing as well as the safety involved for various types of falls.

The displacement of the foot in a longitudinal direction with respect to the ski is harmful during descent and when ski walking. Yet a backward sliding movement of the foot or torsion of the leg is useful in a forward fall.

The horizontal rotation of the foot in both directions—that is, around an axis substantially perpendicular with respect to the plane of the ski—is harmful both in descent and in ski walking but is useful in the case of a forward fall and is absolutely necessary in the case of torsion of the leg with respect to the foot.

The longitudinal oscillation—that is, around a horizontal axis perpendicular to the longitudinal development of the ski—of the foot about the support metatarsus first phalanx is useful during descent, is needed during ski walking, is necessary in the case of a forward fall, and is useful in the case of torsion.

The transverse oscillation of the shin bone—that is, around an axis substantially parallel to the longitudinal development of the ski—is very dangerous both in descent and in ski walking, whereas it is quite indifferent in the case of a forward fall and in torsion of the leg with respect to the foot.

The forward oscillation, longitudinally, of the shin bone—that is, around a horizontal axis perpendicular to the longitudinal development of the ski passing through the malleolus—is useful in the case of descent, necessary in walking, necessary in the case of a forward fall, and useful in the case of torsion.

The backward longitudinal oscillation of the shin bone is not useful in the case of descent, is hardly useful in walking, and is practically indifferent in the case of a forward fall and torsion.

The displacement of the front part of the sole of the foot from the ski is harmful in the case of descent and walking but is useful in the case of a fall forward and necessary in the case of torsion.

As for the eventuality of a backward fall, all the mentioned movements are practically indifferent, except for the backward longitudinal oscillation of the shin bone, which is quite useless.

This invention eliminates the difference between racing skiing, ski touring, and mountain skiing; eliminates danger to the skier's legs; and allows—with soft boots—greater participation in skiing all over the world. These ski bindings are mechanically very simple, and their production cost is only about $10 per pair. Novices can learn much more easily to ski in all conditions, and we can also expect great performances in mountain skiing, which were until now impossible with the present equipment.

The Crab Nebula, the first galactic object to be identified as a radio source.

SEARCHING FOR EXTRATERRESTRIAL INTELLIGENCE

As technology pushes our exploration of the universe ever further, it has become increasingly impossible for most scientists and laypersons to believe that our small planet is the only home of what we call intelligence. Such overwhelming odds are against this situation that few mathematicians could comfortably state that earth is the exception that proves the rule.

Assuredly, we have found no evidence of alien intelligence, but that may be far more the result of our searching methods than of any scarcity of signals. (Radio waves, after all, were not exactly invented; they just needed to be found, having been around for a bit longer than we have.) The quest for some sign of alien intelligence has been carried out mainly in the spare time available for people and for the equipment technically capable of pushing back the boundaries. In this project, a more concerted effort rests on some plausible, scientific assumptions. If the effort were to succeed, the impact on our future could be enormous.

RICHARD M. ARNOLD
3420 Kitzmiller Road
New Albany, Ohio 43054
United States of America
Born December 25, 1944. American.

As a member of the technical staff at Bell Telephone Laboratories in Columbus, Ohio, Richard M. Arnold's responsibilities include designing and developing several types of high-frequency coaxial switches to protect service on a new high-capacity digital transmission system, evaluating and improving reliability of sealed reed contacts used in the switches, and assisting Western Electric Company in the manufacture of these products. His occupation in the area of transmission of radio frequency signals and high-frequency electronics is directly applicable to the project he proposes here.

He brings to his work and his project an educational background that includes a B.Sc. (1966), an M.S. (1968), and a D.Sc. (1970), all taken in electrical engineering from Washington University in St. Louis. He is also the author of several papers on waveguide transmission phenomena and devices, optical detectors, and transmission line analysis.

The project is a radio search for extraterrestrial intelligence. Specifically, we propose to search the sky for narrow-band radio signals of extraterrestrial origin. Such signals could originate from previously undetected natural phenomena or could be intentionally radiated by another civilization within our galaxy. The proposed search would cover a wider portion of the sky, would use more sensitive receiving equipment, and would survey more of the frequency spectrum considered most likely for use by extraterrestrial intelligence than any previous study.

The case for the possible existence of extraterrestrial life is well documented. As the US National Academy of Sciences has pointed out in its recommendations for astronomy and astrophysics for the 1970s, "Our civilization is within reach of one of the greatest steps in its evolution: knowledge of the existence, nature, and activities of independent civilizations in space." The first joint Soviet–American conference on the subject, held at Byurakan, Armenia (USSR) in 1971, concluded, "The striking discoveries of recent years in the fields of astronomy, biology, computer science, and radiophysics have transferred some of the problems of extraterrestrial civilizations and their detection from the realm of speculation to a new realm of experiment and observation."

Previous searches have been relatively limited in either frequency, sensitivity, or area searched. We propose to apply one of the major radio telescopes in the world, the Ohio State–Ohio Wesleyan Radio Observatory, to this problem on a full-time basis. In addition to being able to cover a major fraction of the sky visible in the Northern Hemisphere, this telescope can search the galactic center, an area not accessible to the radio telescopes used in previous searches.

The correct frequency on which to look for an extraterrestrial communication signal is as difficult to predict as the impact on humanity of successful signal detection. Although earlier searches of this type have primarily been confined to the region of 1420.4 megahertz (MHz), an emission frequency of the neutral hydrogen that virtually fills the galaxy, other "natural" frequencies have been proposed. Because emission lines of hydrogen (at 1420.4 MHz) and hydroxyl (at 1667.4 MHz) occupy a relatively quiet region of the radio spectrum and because hydrogen and hydroxyl are the constituents of water on which life on earth (and possibly elsewhere) is based, the region between them has been named the "water hole" and has been suggested as a natural area to search. Thus, in addition to a survey of the hydrogen region of the radio spectrum, we are proposing to search for the first time the upper end of the water hole frequency band, covering the 1652.4-MHz frequency (proposed as optimal because it is based on the geometry of the water molecule), as well as the principal hydroxyl emission frequencies.

Planning for this project is already well advanced. A computer-controlled scanning receiver has been constructed and tested; by itself, this apparatus could be used to carry out a low-sensitivity search. Because our project must rely heavily on volunteers, a monetary grant would not be spent on salaries but on the purchase of a low-noise, broad-band preamplifier that is not obtainable without financial assistance. Such a state-of-the-art amplifier would increase the sensitivity of our receiver by more than a factor of 5, thus increasing its range to almost 2.5 times its present value. This increase in range would multiply the number of potential sites for extraterrestrial life accessible to the receiver, and thus the overall probability of success, by more than a factor of 12.

Because either successful detection of another civilization in space or absolute failure after an exhaustive search would have profound scientific and philosophic implications for humanity, every effort should be made to detect possible signals. The acquisition of the equipment for our search would be a significant step in this effort. Nothing less than the question of our place in space and time is at stake.

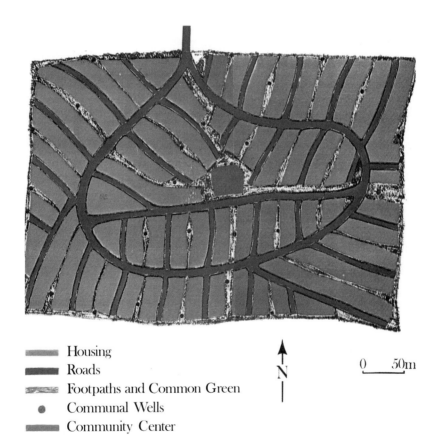

Housing
Roads
Footpaths and Common Green
Communal Wells
Community Center

SELF-HELP HOUSING

Too many social decisions are made by bureaucrats far removed from the specific problems faced by local groups of people. Slowly (most would say "too slowly"), various thoughtful groups are realizing that the best solution to local problems may be to give local people sufficient aid to let them get on with the task of creating solutions best suited to their own needs, interests, and satisfactions. This may hold true for housing the booming populations whose presence in and around established cities poses real problems for Third World governments unable to cope with adequate housing programs.

The so-called squatters of the world, whose shanty towns ring many cities, need a different kind of attention from that normally provided by vast, welfare-oriented undertakings. Shelter is a requirement, and to give it the slightest chance of ever becoming valuable to the community there should be some kind of official recognition of property rights that will encourage the development of shelters beyond their barest requirements. To accomplish this goal requires a kind of housing capable of growing with an owner's financial capabilities and pride of ownership.

Such an approach is the subject of this project. The entrant suggests an attempt to learn how to use the vitality of the squatters for improving their communities.

JULIAN CECIL BOCKS
Honorable Mention, Rolex Awards for Enterprise
5 Thurleigh Court
Nightingale Lane
Clapham South
London SW 12
England
Born June 19, 1905. British.

A member of the Associate Chartered Certified Accountants since 1955 and of the Associate Chartered Institute of Secretaries since 1960, Julian C. Bocks is the administrative secretary of Development Interplan, a group of experts from different disciplines who are involved in the project he describes here.

I propose a scheme for aided self-help housing, with flexibility as an integral part of the design. This scheme is planned for housing problems of the Third World, with particular reference to Sri Lanka.

The proper housing of people with very low incomes is a problem of pressing and growing concern in many parts of the world. The project presented here seeks a practical solution to the problem. It provides an inexpensive but *flexible* design capable of keeping pace with a community's growing demands and aspirations.

The problem can be solved by applying the methods that authorities use to define "acceptable standards" and by providing or regulating housing to meet such standards. Certain advanced economies attempt to bridge the gap between wages and rents by subsidies or allowances, or by grants for building construction. But even advanced countries demonstrate that these efforts in public housing are not always successful, because the standards set are too unrealistic or idealistic for these socioeconomic groups.

In developing countries, the situation is worse because of adverse economic conditions. Here low-income groups are often committed to standards they do not desire, to locations far from their work, and to high rents. As their financial situation is further aggravated, they attempt a makeshift solution by crowding onto idle land around the city fringes—into shanty towns girding the fringes of the actual city—and a population of squatters emerges to bring newer and more urgent problems.

Ad interim, contemporary concepts have increasingly recognized that squatters can be metamorphosed to advantage merely by aiding them to solve their own problems through their own efforts. Thus, the authorities do not need to overstretch housing allocations by trying to build expensive houses of elaborate design for lower-income groups. The principal criteria required for rehousing squatters in any successful housing policy are threefold: security of tenure, where householders have no fear of eviction; location in close proximity to employment; and flexibility of design, so housing can continually change to meet the aspirations of householders and change in their economic state and increasing material opportunities.

Parameters and Policy

To solve the conditions of low-income groups, city housing authorities must:

1. Actively sponsor and publicize a policy of rehousing squatters in houses built by themselves

2. Allow locations, with legal safeguards for a nominal fee, to be selected by the prospective tenants, with reasonable access to the metropolis or large townships by public transport

3. Provision of technical guidance in building these houses in compliance with public health and safety standards

4. Further expansion to be taken into account when considering existing shelter requirements

5. Provision of cash grants (or loans) to aid the purchase of raw materials for provisional shelter while the main building is in progress

6. Maximize the use of locally available materials

7. Utilize the labor of these groups, except for specialist work

Aided Self-Help System

The system has been created to satisfy the social and cultural requirements of the lower socioeconomic groups. It consists of three complementary optional designs for single-story, low-cost houses designed to satisfy different requirements. Type 1 is the most economic, Type 2 the most flexible for further expansion, including the capacity for a mezzanine floor. Type 3 is capable of expansion even to the standards of the middle socioeconomic group; that is, with attached bathroom, three bedrooms, and a back veranda.

Flexibility and choice have been incorporated in the design to enable future expansion to be carried out as the dweller's ability to invest in more space materializes. The houses can be constructed by relatively unskilled prospective tenants under the supervision of housing authorities. Only low-cost materials that are readily available locally have been specified.

The basic housing unit (Type 1) is designed to satisfy the needs of lower socioeconomic groups with incomes less than $350 per month. The design parameters have been set to satisfy the following urban physical conditions: single units (bungalows), dual units (semidetached), or "series" (terraced).

The foregoing recommendations form the nucleus of a research project presently being carried out by professionals in a group called Development Interplan. The group is composed of specialists in the various necessary disciplines.

The Department of National Housing in Sri Lanka has expressed great interest in the scheme and has forwarded us their brief. They have reserved an actual site in the greater Colombo area for the pioneering of this scheme. Our undertaking is purely on a voluntary basis, the only understanding being that, on the implementation of the research so far achieved, they will keep us informed of progress. This feedback information will enable deductions in relation to the further development of this philosophy of solving housing problems. These deductions will, in turn, provide a knowledge base for the application of this principle in other parts of the Third World.

"SHAKE-PROOF" CAMERAS, BINOCULARS, AND TELESCOPES

How many times have you had a photo come back slightly blurred, not because your camera was improperly focused, but because the available light, lens, film speed, and lack of a tripod forced you to hand-hold the camera at a shutter speed where stability declines? Or, with telescope or binoculars have you encountered the frustration of trying to keep the distant objects centered?

No matter how proud we may be of our practiced "iron grip," there is a point at which the tiniest tremors of pulse rates or outside vibrations cause us to lose the picture. Wouldn't it be nice to forget this irritating problem forever? If the project outlined here works out as planned, we may be able to do just that.

ADRIAN ANTHONY CECIL MARCH
High Beeches, Swaineshill
Alton, Hampshire GU34 4DP
England
Born December 11, 1931. British.

After earning his school certificate and higher school certificate in 1945–50 at the Bryanston School, Adrian March won a major scholarship to Sidney Sussex College, Cambridge, where he was a senior scholar and took his M.A. in natural sciences and electrical engineering in 1953. Between 1953 and 1956, he researched microdensitometry and the structures of thin films and took his Ph.D. at the engineering laboratory at Cambridge University.

As a consultant engineer, under the company name of Adrian March Research Ltd., his principal fields of activity are optics, electronics, and precision mechanisms, with a specialty in aircraft equipment.

I investigated the feasibility of designing lens systems suitable for hand-held optical instruments (such as binoculars, telescopes, and cameras). The image would be stabilized against inadvertent angular motion of the instrument by means of a passive assembly of refracting elements on a suspension that isolates them from angular disturbances. This system allows, for example, the construction of a "shake-proof" camera.

Optical sight-line stabilization techniques are well known in observations at relatively high magnifications of moving vehicles, aircraft, or ships. Such techniques almost invariably require a gyroscopic directional reference and a servo system that activates a reflecting or refracting optical element in an appropriate relationship to the angular deviation of the instrument. In addition to requiring a source of power, this kind of stabilization system adds considerably to the size, weight, and cost of the instrument. Such arrangements are necessary when, for example, precision stabilization is required against the motion of an aircraft over a large angular range, but they tend to be both impractical and uneconomical for cameras and binoculars, which are mainly subjected to small disturbances caused by vibration or muscle tremor.

It appears possible to design lenses in which the image is stabilized against small angular disturbances by mounting some elements on a suspension system having an axis isolated from such disturbances. An arrangement of this type, consuming no power and imposing no more than a minimal size and weight penalty, opens the door to the concept of "self-stabilizing" binoculars and camera lenses.

For the purpose of illustration, I will describe the principle by referring to a simple telephoto lens, but the description could equally well apply to a camera lens or to the objective of a telescope or monocular.

In illustrations, it is often difficult to visualize the operation of a sight-line stabilizer when the visible world is drawn as fixed in relation to the paper. It is usually much simpler to study the operation by deviating through the same angle both the entering rays and any component whose direction is stabilized in space. This approach has been adopted in the illustrations that follow.

Consider a basic telephoto objective (Figure 1). The image can be stabilized by moving the two lens elements in such a way that the principal focus of the positive element lies always on the line XY joining the focal point, X, to the center of the negative element (angles and displacements are shown greatly exaggerated). See Figure 2.

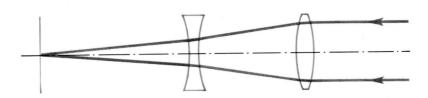

Figure 1.

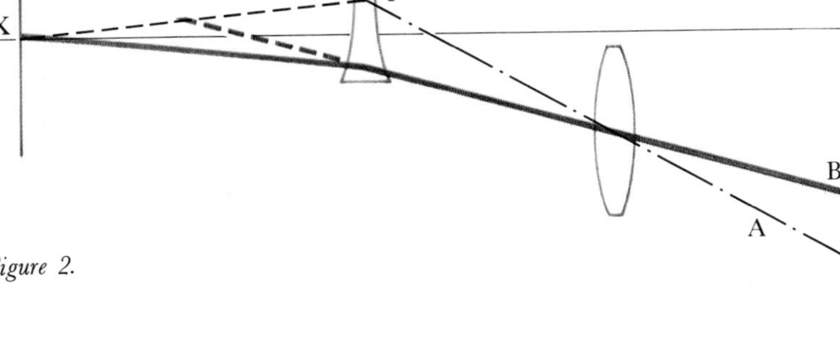

Figure 2.

The line, A, joining the centers of the two elements must move through a greater angle than the entering ray, B, so the scheme as it stands is unsuitable for passive stabilization.

Now consider an equivalent system formed by replacing each element with a retrofocus combination. In Figure 3, the physical components are drawn in solid line, and their optical resultants are shown as dotted outlines. If the shaded elements are displaced laterally, the line joining the centers of the (dotted) resultants can be arranged to move through a greater angle than the line joining the centers of the shaded elements. By this means, effective angular movement can stabilize the image, although the axis of the physical elements remains parallel to the entering ray. See Figure 4.

It therefore becomes feasible to design a system in which sight-line stabilization is achieved by stabilizing the axis of the shaded elements. Provided that only small angles are concerned, it is considered probable that their principal planes need not remain precisely parallel and that they can be rigidly connected to form a single unit, which we can refer to as the *stabilizing unit*.

A completely passive self-stabilizing lens theoretically could be produced by mounting the stabilizing unit on two pairs of low-friction bearings so inertia maintains its axis, and hence the sight line is stabilized in direction. In practice, a small restoring force is desirable to prevent the stabilizing unit from abruptly reaching its limits of movement and to permit steering the sight line by moving the instrument. This, in turn, necessitates some form of damping to prevent oscillation.

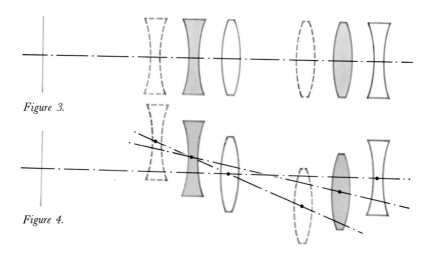

Figure 3.

Figure 4.

To establish the principle, it is of little consequence whether the lens in question is for a film or television camera, binoculars, or telescope. I propose, therefore, to facilitate recording and analysis of performance by choosing a lens equally suitable in observation for static measurements and in a television camera tube for measurement under dynamic conditions.

The optical part of the project will accordingly require the design of a lens of approximately 180 millimeters focal length and 45 mm aperture. The effect on the displacement aberrations of the stabilizing elements will be examined and a relationship established between optical resolution and sight-line angular displacement for the two alternatives of parallel and rotational displacement. The object will be to determine the simplest design that will maintain adequate resolution over an angular range of the order of $\pm 2°$.

Having confirmed the approximate values of the masses and moments of inertia of the moving elements, I can establish the optimal design of the suspension on the basis of electrical wave filter theory by treating the system as a low-pass filter having a cutoff frequency of about 1 hertz.

Then a prototype of the design will be constructed and tested to confirm the validity of the theory and to evaluate the practical performance of the system. Detailed records will be kept of the theoretical and practical work, and the project will be concluded by an assessment of the most probable fields of application for the technique.

SAVING THE ABYSSINIAN WOLF

At the time he conceived this winning proposal, the entrant had already made a preliminary field trip to the high plateaus of Ethiopia, home of the Abyssinian wolf. Preparing for a return expedition aimed at rescuing this animal from near-certain extinction took time, and in that time events moved swiftly against the threatened wolves. First, political upheaval in Ethiopia upset carefully laid plans with concerned local governmental authorities and conservation officials. Second, and even worse, a war between Ethiopia and Somalia raged in the general area of the wolves' habitat. The urgency of this project has thus been heightened by humans, who are both the greatest nemesis of this sociable species and its possible saviors.

KENNETH LEE MARTEN
Rolex Laureate, Rolex Awards for Enterprise
Department of Zoology
University of California
Berkeley, California 94720
United States of America
Born January 5, 1946. American.

As a doctoral candidate in the department of zoology at the University of California at Berkeley, Kenneth L. Marten is pursuing the same field, zoology, in which he took his B.A. from the same university. He has been a part-time teaching assistant for courses in animal behavior, animal evolution and diversity, neurobiology, and introductory biology.

Marten has had extensive experience in biological field work, involving the live trapping and handling of carnivores and rodents, immobilizing monkeys, radio tracking, and behavioral observations, often in high-elevation mountainous habitats and the tropics. He was a Peace Corps volunteer in Ghana, has traveled widely through Africa, Latin America, Europe, and North America, and has sailed and navigated a small boat from California to Hawaii and elsewhere in the Pacific. He speaks French, German, Spanish, Twi, Fante, and Brusa (a West African dialect).

In his project, which he describes here, Marten will be joined by his wife, Linda Hobbet, who is an artist and a specialist on canines (her drawing of two Abyssinian wolf pups is reproduced on page 258).

My wife and I plan to develop and implement a management program for the endangered Abyssinian wolf (*Canis simensis*) in Ethiopia. More information and immediate action are needed to conserve this little-known animal. We will interview the inhabitants of the Simien Mountains National Park to learn how many wolves are killed by people per year, if wolves kill livestock, and what could persuade people to stop killing wolves. We will also examine wolf scats (droppings) to document possible wolf predation or scavenging of livestock and will observe wolves to determine if deaths caused by humans exceed the births.

In the Bale Mountains, where wolves are relatively numerous, we will study the factors that limit their populations: food supply, habitat preference, mortality, and social behavior. In collaboration with the Wildlife Conservation Organization of the Ethiopian government, we hope to train Ethiopians to continue our studies, and we will help them develop a practical, step-by-step recovery plan for the wolf. This program will begin during the study and continue after we leave.

The Abyssinian wolf (also known as the Simien fox) is a handsome, little-known carnivore that has suffered from human activities. Now, unless their old enemy lends a hand, the wolves face almost certain extinction. They are presently restricted to two remote mountain plateaus in Ethiopia. Our information about them was gathered in my own ten-day pilot study (1975) in the Simien Mountains and in a six-week census (1976) by James Malcolm of Harvard University of the wolves in the Bale Mountains (which have a larger, more viable scientific community, as well as to all people concerned about preserving wildlife and wild places on this planet. Such studies would not be difficult, as the habitat is open and as the wolves are diurnal and approachable (photographs taken by Dr. Richard Wrangham on a one-day visit to Bale dramatically illustrate these points). Moreover, studies could help shed light on several current issues in sociobiology, such as the interaction of ecology and social systems (the wolves appear to have a social system that may be especially sensitive to ecological parameters), and kin selection and the evolution of altruism. The interests of conservation and science are thus mutually supportive.

J. G. Stephenson, of the Ethiopian Wildlife Conservation Organization, has emphasized the urgent need for detailed information about the wolf's ecology and way of life as the basis for developing a management policy. An attempt is being made to establish a park in the Bale Mountains that will encompass the current range of the wolf. Stephenson has also noted that a conservation- and science-oriented study could significantly advance this goal.

We intend to begin by systematically collecting and analyzing data in two areas: one with many wolves and one with few. To gather information on the predator-prey system, we will:

1. Take a census of wolf and prey populations every three months

2. Determine the percentage by weight of each prey species in the wolf's diet

3. Determine natality and average litter size and estimate adult and juvenile mortality

4. Analyze wolf and prey densities through the seasons to determine the effects of prey density on wolf hunting behavior, population, and reproductive success

5. Compare results from both areas to determine the effect of wolf (and probably prey) density on predation parameters, such as percent of time foraging, diet, and seasonal changes in population size.

To collect data on the wolves' social system, we will:

1. Determine if wolves live in larger groups and are more social at high density than at low density

2. Count the incidence of helping at high and low density and assess its demographic causes and possible ecological correlates

3. Follow juveniles to see where young wolves settle and how successfully they survive and reproduce, particularly with respect to human-caused barriers such as a new road

In addition to the detailed work in Bale, we will visit the Simien Mountains National Park (if and when politics permit), where a small population of 15 to 20 wolves lives in conflict with pastoral natives of the area. By interviewing inhabitants of the park, we hope to determine if they kill wolves, how they do it, and roughly how many they kill per year, to obtain information about wolves killing livestock, and to find out what could persuade people to stop killing wolves. We will also estimate natality and determine if human cropping of the wolves is responsible for their low numbers. And, finally, we will investigate juvenile mortality and nonhuman sources of mortality.

With officials of the Ethiopian Wildlife Conservation Organization, we hope to derive and implement a recovery plan, which will be improved as we obtain more information about the wolf and its problems. The plan and its application will thus fuse scientific research and administrative practicality.

The "recovery plan" is a method recently adopted by the United States Fish and Wildlife Service to manage endangered species, such as the red wolf (*Canis rufus*) and the California condor. The basic method is to bring together biologists, government officials, and (when necessary or appropriate) private parties, to develop a cooperative, practical management program to save a given species. They draw up a step-by-step plan almost as detailed as a cookbook or a flow diagram. As alternatives are encountered, decisions based on biological information are made. While we are in Ethiopia, we want to write and begin a recovery plan for the Abyssinian wolf as soon as enough information is available and to update it continuously until our departure (we may have some additional contributions to make to this program after further analysis in the United States). We will also help train Ethiopians to carry on the program after we are gone.

Since we began the search for funds to do this project, the political situation in Ethiopia appears to have deteriorated. However, a detailed report from James Malcolm, who was in the Bale Mountains for five weeks in January and February of 1977, suggests that it is safe to do field work there. We therefore hope to proceed as soon as the situation in the Ethiopian capital, Addis Ababa, improves.

THE GOLD HAS BEEN THERE FOR 270 YEARS, AND NOW...

Gold and silver. Worth an estimated $100–200 million. Just sitting there. Documented. No doubt about it. It all went there, if you want the exact date, on October 22, 1702. Hundreds of witnesses. Repeat, no doubt about it.

If that scratches *your* enterprise itch, don't feel alone. Over 50 attempts have been made since then to get it, some with a bit of success.

You'd better hurry, though. There's a project afoot, using some very fancy equipment, that may get there first.

CHARLES PLACIDE GUICHARD
21130 Wardell Road
Saratoga, California 95070
United States of America
Born November 29, 1919. American.

A B.A. in 1941 from the University of California at Berkeley, an M.A. in 1948 from the University of Southern California, and a Ph.D. from Stanford University in 1957 form the professional background of Charles P. Guichard, who is now a teacher and educational consultant in Sunnyvale, California. He has 30 years of teaching, research, and administrative experience in a variety of fields at the high school, college, and university levels. He has taught courses in psychology, history, and product engineering and evaluation. He has conducted research in tests for pilots, navigators, and bombardiers; in motion picture audiences; in personnel administration; and in racial stereotyping.

During the War of Spanish Succession, on October 22, 1702, a British fleet commanded by Admiral Rooke surprised a Spanish treasure fleet under Admiral Velasco in Vigo Bay (Galicia, Spain). Admiral Velasco, seeing his ships falling into English hands, ordered them set on fire. Seventeen galleons burned to the water line and sank, taking their treasure with them. Although estimates vary, the value of the treasure was probably between $100 and $200 million, by modern standards.

Since 1702, 50 or more salvage attempts have been made. The following have met with partial success: Rivero (1748), Evans (1772), Dickson (1825), Magen-Bertge (1869), Gowen (1885), Pino-Iberti (1903-28), van Wienen (1939), Potter (1955-57), and Stenult (1958). Each increase in scientific knowledge and improvement in the arts of diving, salvage, or metal location has stimulated a fresh attack on the problem. The history of Vigo Bay salvage over the past 250 years has paralleled the development of salvage methods.

I propose to locate and salvage the treasure. A team of specialists, directed by Dr. Broussal, Mr. McGill, and me, will locate some of the sunken ships and salvage cargo by means of the techniques outlined below. McGill will develop new sensor techniques for exact location of the hulks, I will make a historical analysis of the team's ten most significant past expeditions, and Broussal will render my information into navigational charts and will supervise the small boat and diving operations.

The most promising technique for locating the Vigo Bay galleons, now buried under a layer of mud approximately 5 to 10 meters deep, is offered by a synthetic aperture sonar device. A combined acoustic transmitter and receiver, mounted on the hull of a small boat, is moved in a straight line over the area to be searched. The sonar signal is processed to reconstruct a two-dimensional image of the spatial position of all large acoustic scatterers in the area.

The synthetic aperture technique is usually referred to as *side-looking sonar*, as it is usually employed to search a horizontal area on each side of a ship's path. In this case, the reconstructed two-dimensional image is in a horizontal plane. However, by using vertically oriented acoustic transducers with high directionality, the same technique can be used in the vertical plane to produce an image in depth, even for objects buried 7 meters in mud.

Commercially available side-looking sonars, possibly with modified acoustic transducers, would be suitable for vertical soundings of mud. A shore-based helium–neon laser would allow precise positioning of the sonar-carrying boat.

To our knowledge, synthetic aperture sonar for vertical soundings of mud has not been used in any treasure recovery operation, at Vigo Bay or elsewhere. I hope this technique will prove useful in treasure-hunting operations where the area of mud to be searched is too large for conventional coring techniques.

I plan to start the project in July 1977 and will spend three months on exploratory work, mapping, and equipment modification. If the technique is successful for location and salvage on a pilot basis, I plan to continue the search for three years.

Elephas maximus, *the Asian elephant.*

ELEPHANT RANGES, WHERE PEOPLE AND ANIMALS COEXIST

Oddly enough, in an era when it seems that most of our works are detrimental to wildlife, there are certain specific situations where our development of the environment can actually enhance the living conditions of certain species. The Asian elephant appears to be such a species, and recent research findings may be of considerable importance to the future of both people and animals in Asian countries.

Once ranging from Iran to northeastern China and on down to the tip of Malaysia, the Asian elephant today has dwindled to a population that inhabits much more limited pockets of territory and faces increasing problems as the result. Humans have done the pocketing in our insatiable search for more space and resources, and the resulting pressure on the elephants has constricted their natural migratory and ranging patterns. This, in turn, leads to the elephants' frequent disruption of agriculture, property, and even human life. Now, given a clearer understanding of the elephants' needs and of the fact that they can forage more effectively in logged forests, where secondary growth supports more pachyderms than do virgin forests, a move is afoot to manage a more equitable arrangement between people and the great beasts who have lived in Asia for centuries.

ROBERT CHARLES DACRES OLIVIER
Honorable Mention, Rolex Awards for Enterprise
Department of Applied Biology
Cambridge University
Pembroke Street
Cambridge CB2 3DX
England
Born March 5, 1950. British.

At the time he submitted his project, describing himself as a fifth-year research student at Cambridge University's department of applied biology, Robert Olivier was preparing his Ph.D. thesis (entitled "On the ecology and behaviour of the Asian Elephant (Elephas maximus) with particular reference to peninsular Malaysia and Sri Lanka"). In addition, he was co-chairing the Survival Service Commission of the International Union for the Conservation of Nature and Natural Resources' new "Asian Elephant Specialist Group" for 1975-78. This position involves coordinating the group's conservation activities in a dozen countries and managing an international Asian elephant secretariat from his Cambridge laboratory.

Since taking his First Class B.Sc. (honors) degree in zoology from the University of Bristol in 1972, Olivier has studied rainforest primates in Malaya with a Cambridge University expedition. He spent three years in Malaysia (as an honorary research officer of the federal game department), six months in Sri Lanka (as project leader of the Smithsonian Ceylon elephant ecology project), and made study tours in southern India, Nepal, Assam, Burma, Thailand, and Sumatra. He is also a licensed pilot, which is a key consideration in the techniques he discusses in his project.

Asia has lagged behind the rest of the world in conservation and management of wildlife and in the necessary research. This situation has not arisen out of a complacency born of large numbers of remaining animals. On the contrary, the massive human population explosion in the region has resulted in the withdrawal of wildlife to more hostile or inaccessible areas, away from the minds of most authorities and people. As a result, Asian wildlife is not nearly as abundant as African wildlife and has not become the basis of an important industry. Moreover, conservation research has not been encouraged by Asian governments and is not facilitated by Asian habitat types.

In view of these circumstances and of the inroads into remaining natural habitats that are currently escalating exponentially, the approaching doom of Asia's wildlife and forests has become a matter of grave concern to conservationists. The African approach of setting aside large areas solely for wildlife is politically unfeasible in Asia.

Our present knowledge of Asian mammal ecology indicates that they generally prefer habitat types influenced by fire and flooding; these have been the first to disappear under the tide of human influence. Prior to human intrusion, the mammals must have been living in such regions in much higher densities than today. They now have been forced into the more sterile habitats. Even these habitats, however, given improved technologies and the demands of the industrial world, are now being developed and exploited.

I was recruited by the Malaysian government to study the ecology and behavior of elephants, which—as a result of range depletion—often cause serious damage in agricultural programs. Some interesting ecological facts emerged from the study.

In brief, elephants on primary, virgin forest ground were found to require a range five times larger than those living in disturbed, secondary forest that had been selectively logged. Objective measures revealed much more available food in the secondary forest, thus explaining the range difference. Recent research has likewise confirmed a preference for secondary habitats in the totally different southeastern Sri Lanka dry zone, where such ranges are often created by artificial reservoirs as well as by the limited forestry.

Thus certain human activities on these sterile, or forgotten, remaining natural habitats in Asia apparently lead to an artificial, human-induced mimicry of the ecologically preferred habitat types that have largely disappeared. A review of our knowledge suggests that this principle applies to most of the larger animals, not just to the elephant.

It is ironic that in its initial stages the very process of land development that ultimately threatens the survival of these animals actually can improve their living conditions. But therein lies the only hope for conserving most of the larger mammals that require much space and also for minimizing their conflict with humans in the overpopulated Asian region. This hope lies in managing habitats to allow limited human exploitation that would not adversely affect, and perhaps would even improve, wildlife habitats and that would not prevent migrations or fragment populations. Of course, special reserves will always be needed for plants and the more sensitive animal species. I feel we are on the threshold of a breakthrough in Asian conservation, for here we see a way to compromise on the apparently unacceptable demand for large land areas for animals.

I hope to develop and apply these ideas, particularly in Sri Lanka and India. The idea has been well received and encouraged by Asian authorities, and it is to this end that my project is directed.

A Managed Elephant Range in Sri Lanka

Because it has the greatest known natural range of the species concerned, the elephant is the first animal to come into conflict with people. Mutual suffering, owing to indiscriminate land development, ensues. The need to protect people and elephants from each other is indicative of larger issues of wise land use, proper planning, protection of catchment areas, soil and water conservation, stability of local meteorological patterns, conservation of other species, and so on. All these things are favored merely by a consideration of the elephants and their ecology. Thus the elephants' welfare becomes an indicator or baseline against which to assess the impact of people on the environment.

In Sri Lanka, following government policies designed to achieve self-sufficiency in food as soon as possible, the rapid pace of the development programs (which, in common with most such programs carried out or planned in Asia, gave absolutely no consideration in the planning stage to the effects on wildlife) has already created a situation in which elephant herds have become pocketed in several areas and constitute a real threat to human produce, life, and property. However, in view of current proposals to implement over 20 massive irrigation projects between 1977 and 1982, the current situation represents only the tip of an iceberg. The potential conflict between people and wildlife in the wake of these projects, coinciding as they do with the largest remaining population of elephants, is considerable and is detrimental to both. This serious situation—which, once created, will be extremely hard to resolve—will materialize unless action is taken soon.

I propose the establishment of large sanctuaries to protect the full seasonal ranges and movements of the remaining (relatively large) elephant populations. According to our ecological knowledge, this measure should halt imminent escalation of the pocketed herd phenomenon, which ultimately threatens the survival of the elephant (and other species) in free-ranging populations. The idea is to identify large elephant populations and to develop elephant ranges as managed reserves. This has been done for Sri Lanka and includes a large area in the southeast, incorporating two national parks and three wildlife sanctuaries. The precise boundaries would be based on inclusion, if possible, of *all* areas (as revealed by radio tracking) visited by elephant groups whose ranges overlap at one time or another. Whatever of this large area can be incorporated creates a zone around the national parks and other inviolate corridors and can act as a buffer for them. The buffer zone then constitutes the range, and the whole comes under the active control of the appropriate governmental department. Outside the parks and inviolate areas, a variegated land use approach would be taken within the range based on present ecological understanding. The following would be permitted, subject to control according to research recommendations: human habitation, shifting cultivation, livestock grazing, fishing, hunting, and selective logging.

The matter requires considerable research not only into elephant seasonal ranges but also into other wildlife, habitat productivity, and appropriate development proposals for the area. I list here some of the special techniques required.

1. Aerial mapping of wildlife areas with potential for the application of this concept. This requires assessing the continuity of forest cover, vegetation mapping, nature and extent of human activity in the area.

2. In suitable areas, regular long-term aerial monitoring of animal numbers and of the distribution of all visible species.

3. Temporary chemical immobilizing of indicator species (in this case, the elephants) and attaching radio-transmitter collars. Animals would be collared in various parts of the range during the wet season to see where they go during the dry seasons, and vice versa.

4. Long-term tracking of collared animals by using a light aircraft with receiving equipment to locate position and seasonal ranges, habitat preferences, and associated daily and seasonal requirements.

5. A good deal of on-ground work to complement the preceding, to acquire data to interpret findings, and to set levels on human activities within the elephant range. These do not qualify as special techniques and may be carried out by other personnel associated with the project.

I have applied for the Rolex Award in the hope that it can be used to contribute largely to the purchase of the aircraft so vital to the surveying, radio-telemetry research, and travel inherent in this project. This equipment is not yet available in the Asian region and has yet to be developed there. I hope that one result of the wider project will be to train Asians in these techniques. In that event, it might be possible to hand over the aircraft to suitable government or university authorities for continued research and monitoring. Aircraft are also of great value to wildlife sanctuary personnel in law enforcement activities.

My experience with aircraft for wildlife research began in 1969 in Africa. Then I decided to try developing the techniques in Malaysia, where I temporarily immobilized and collared several elephants, whose movements were later followed from the air over a long period. I have also been involved in aerial tracking of collared tapirs and gharials.

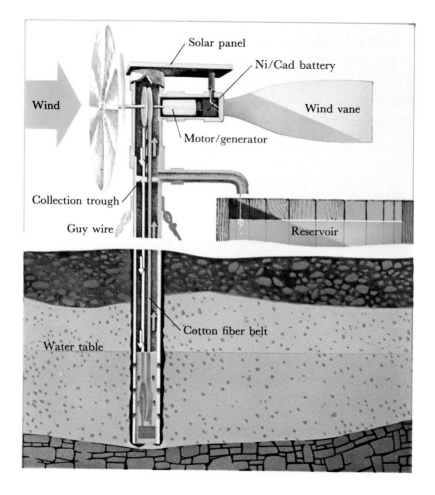

A SOLAR/WIND WATER PUMP

In those parts of the world where water is scarce, efforts required to lift it from subterranean sources consume much time and scarce energy resources. Power is often needed over long periods each day to provide essential water supplies for human and agricultural needs. The benefits of a system for raising water under economical conditions would be enormous. This project offers a potential solution to governments and other organizations concerned with the issue.

MITCHELL HARRIS ZUCKER
P.O. Box 1023
Mendocino, California 95460
United States of America
Born October 30, 1935. American.

After taking his B.S. in industrial engineering from the University of Michigan in 1957, Mitchell H. Zucker became a project engineer in inertial guidance systems with the Ford Instrument Company, Division of Sperry Rand Corporation. Later, with the autonetics division of North American Aircraft Company, he worked on inertial guidance computers before establishing and managing the behavioral research laboratory shops in the psychology department of the University of California at Davis. He is a writer and an illustrator.

I have successfully tested the operating principle for my solar or wind water pump, which can raise hundreds of gallons of water daily at a fraction of what conventional pumps cost. Loss in delivery flow rate is compensated by trouble-free operation at no energy cost. This most efficient water-raising system is ideal for economically depressed arid regions.

Design Principle

A continuous belt of nylon-reinforced cotton fiber saturated with water is rotated either by a pulley and a propeller or by a fractional-horsepower, solar-activated motor. The saturated belt passes around the upper pulley and continuously squeezes water against the pulley. The water collects and drops from the center of the pulley into a trough. A weighted lower pulley provides belt tension and alignment.

At slow speeds (10–60 revolutions per minute, or rpm), water drips steadily from the upper pulley (about 1 gallon per hour). As the speed increases (60–150 rpm), a steady stream results (about 4 gallons per hour). At greater speeds (150–300 rpm), water is continuously sprayed around the entire pulley area (6–10 gallons per hour).

The design is based on a direct application of the operating principle, using reinforced cotton rope. As research funds become available, I will construct experimental models that use wide-band reinforced cotton belts, synthetic sponge rubber belts, and sprocket drives for maximal water transfer.

Design Data

The belt is $\frac{3}{16}$-inch nylon-reinforced cotton fiber, the pulley diameter is 3 inches, and the pressure head is 30 feet. Each additional 200 rpm yields 0.003 horsepower and a minimum of 6 gallons of water per hour (12 gal/hr with a 1-inch cotton belt). The minimum propeller diameter (based on Betz's theorem) is $1\frac{1}{2}$ feet, and the motor generator produces 12 volts each 180 milliamperes.

Solar panels augmented by rechargeable batteries are available to provide continuous operation, even when there is no sun. When wind-powered rpm exceed motor-powered rpm, an electronic circuit control automatically switches the motor to generator mode for day or night battery charging. With no sunlight and a discharged battery, the pump will deliver water at wind speeds of 3–5 miles per hour.

I first designed this pump while living in a village in southern India, where all household water is raised by hand from a large, shallow well located near a runoff source of fecal contamination. My first engineering model, pieced together from found materials, demonstrated that it could profoundly improve the lives of millions of people if made from high-quality materials.

By using the natural filtering properties of soil and by locating small-bore, slow-recovery wells away from contaminated sources, clean household water can be steadily raised at no energy cost. The manual labor installation time for a 20-foot well (after locating the assembled pump near a known source of water) is 3 person-days. This includes hand augering a 6-inch × 20-foot well (clay and gravel), lowering the pump, securing guy wires and tripod mount, testing and alignment, and catching water. The pump is designed to operate continuously 24 hours a day and is expected to require no maintenance for 7000 hours of motor and generator operation.

Prototype field testing will commence August 1, 1977, during drought conditions in northern California. Water will be raised from a 6-inch × 20-foot test well located in an area where large, costly, shallow wells normally exhibit slow recovery rates. The tests will run continuously throughout the driest part of the year and will provide constant water for an animal-watering station. The expected test completion date is November 1, 1977.

Trichechus inunguis, *the Amazonian manatee.*

THE MERMAIDS OF THE AMAZON

The legendary Mermaids may survive forever in myth, with little danger of becoming extinct in our memories and fancies. However, we have shown ourselves capable of totally eradicating the real mermaids, as pointed out in this project designed to prevent a repetition of history. It is painful to know, as we must, that we humans have killed entire species, not just out of careless neglect and disinterest, but through deliberate action. More painful yet is the thought that no individual was able to raise a voice sufficient to prevent the slaughter.

This project is aimed at preventing another such unconscionable extinction of a species. To lose the mermaids of the Amazon would tarnish the memories of legend and of our own world in the eyes of future generations.

HENDRIK NICOLAUS HOECK
Blasenbergstrasse, 7
CH 6300 Zug
Switzerland
Born January 1, 1944. Colombian.

After studies in Colombia, Hendrik N. Hoeck went on to Bonn University in Germany. He earned his diploma in biology from Munich University in 1970 and then entered the zoology department of the university and the Max Planck Institute fur Verhaltensphysiologie. His work there took him into field research in Tanzania, and he completed his Ph.D. in 1975. As an employee of the Max Planck Institute, he has been involved in five years of field observations of the ecology and social behavior of the rock hyrax and bush hyrax in Serengeti National Park, Tanzania.

Mythology has many legends about mermaids, beautiful aquatic creatures with long hair, breasts, and human features. It took humans only 27 years (1941–68) to slaughter to extinction the largest of these mermaids, Steller's seacow, *Hydrodamalis gigas.*

Steller's seacow belonged to the order Sirenia, the largest strictly vegetarian aquatic mammals. Today, we can still find most species (three manatee, one dugong) of this once successful group, but all are on the list of endangered species. Among them, the Amazonian manatee, *Trichechus inunguis,* is probably one of the world's most acutely endangered mammals. One researcher writes, "ruthless hunting for its meat has brought it nearer to extinction than perhaps any other mammal of the Amazon region. It is now so rare that only by the prohibition of all hunting can the species be saved."

Compared with the other manatees and dugongs that live in the sea, *T. inunguis* is especially vulnerable; it inhabits blackwater lakes, lagoons, and waterways in the lower reaches of the main tributaries of the Amazon. Rivers are limited habitats, easily altered or destroyed completely. Given the enormous development and human colonization of the Amazon basin, the Amazonian manatee is certainly doomed if no immediate steps are taken to rescue it.

Our current knowledge of *T. inunguis* is very poor. We know that the manatees have an ecologically important function of clearing rivers and lagoons of vegetation and keeping them open (it has been suggested that they be used for weed clearance of channels and waterways). Also, we know little about their migrating patterns, which seem to depend on the widely varying water level of their homes. During floods, manatees seek inundated jungles and thickets, where they are relatively safe from humans. When the water is low, they must move into rivers, which are more exposed. In extreme drought, the animals may be trapped in shallow pools. One scientist describes how, in 1963, thousands of manatees were trapped in such pools in Brazil, Peru, and Colombia and were slaughtered by local inhabitants, who developed an irrational killing frenzy.

Luckily, there still appears to be just enough time for the preservation of *T. inunguis*. A recent report from Iquitos, Peru, states that several thousand manatees have been sighted in the Samiria, Pacaya, Nanay, Orosa, and Pastaza Rivers, in the north and south of the Amazon River network.

In my proposed project for promoting the conservation of the Amazonian manatee, I will first survey this district to establish the actual stock. The ultimate aim is to suggest one area that can be selected as a conservation area or possibly a national park.

Many national parks have one big fault; they were set up prematurely, without knowledge of the ecology of the animals they were meant to save. The animals cannot read maps and so cross boundaries only to be shot dead. People must first know the animals' own territories and home ranges and must then plan accordingly. To gain knowledge of a manatee population's habits and haunts will be the main object of the project.

A national park, however, will neither come into existence nor survive without wide human interest. Publicity is the key. One must both interest the appropriate organizations and appeal directly to the public. From the beginning, I plan to work in close cooperation with the conservation authorities of Peru. I will be supplying regular reports to several international organizations—the International Union for Conservation of Nature and Natural Resources (IUCN), FAO Advisory Committee on Marine Mammals Research, World Wildlife Fund (WWF), Frankfurt Zoological Society, New York Zoological Society, and others—so at the end of the project its continuation and extension will be encouraged. I also intend to make many photographs for popular articles and films for television.

The work will include several aspects:

1. A survey of the area, mapping the abundance and distribution of *T. inunguis,* must be made.

2. I will choose a focal area that contains an adequate number of animals, is suitable for regular monitoring, and has little human interference.

3. In this area I will attempt to estimate the actual number of animals with at least two counts: one in the dry season and one in the wet season.

4. Individuals of both sexes will be marked (by tags and/or radio collars) to gain information on home ranges and migration.

5. I will gather information on the manatees' activity, population dynamics, feeding ecology, and social organization.

6. Special emphasis will be placed on photographs, films, and sound recordings.

THE FOOLPROOF CAMERA

For the interested amateur or hard-working professional, the dizzying pace of developments in the technical world of photographic equipment makes *obsolescence* a bugaboo word. No sooner does one adopt the latest level of technological wizardry than yet another advance is made in the state of the art. Is there a truly perfect system, an ultimate development or refinement beyond which we need not explore? It's doubtful, as few would deny. Yet suppose there were a system that could put the control of optical beams easily, and even totally, within the hands of the photographer. A boon? Definitely.

It appears that, technically, such a system could be created. The project described in the following pages is aimed at achieving this goal.

THOMAS KEITH GAYLORD
92 26th Street, N.W.
Atlanta, Georgia 30309
United States of America
Born September 22, 1943. American.

After taking his B.S. in physics (1965) and his M.S. in electrical engineering (1967) from the University of Missouri at Rolla, Thomas Gaylord earned his Ph.D. in electrical engineering at Rice University in 1970. As an associate professor in the School of Engineering at Georgia Institute of Technology, his work involves teaching and research in the areas of solid-state electronics, optics, lasers, and holography—an appropriate basis for the project he describes.

I will develop an extremely accurate, modular analog/digital electronic system for controlling an optical shutter. This system will allow remote control of optical beams and will be externally programable. Such a system is not commercially available at present.

Operational Characteristics

A large aperture ($<$ 50 mm), electromechanical, back-to-back double shutter (one normally open and one normally closed) will be controlled by four primary modes of operation: bulb, manual, timer, and exposure. Operation can be initiated manually or through remote programability.

In the bulb mode, the shutter is opened when the actuating pushbutton is depressed and remains open until the pushbutton is released.

In the manual mode, the shutter is opened when the actuating pushbutton is first depressed. It remains open until the "stop" pushbutton is depressed. In both bulb and manual modes, the elapsed times are continuously displayed on a digital readout.

In the timer mode, the shutter is opened for a preset time that is continuously selectable from 1 millisecond to 10,000 seconds. The timing error does not exceed 0.05 percent of the set value.

In the exposure mode, the optical power (in watts) is continuously measured and integrated to give the exposure in joules/meter^2. Exposures from 0.1 joule/meter^2 to 10^6 joules/meter^2 are measurable to within 0.05 percent of the preset value. There are two digital readouts, one continuously displaying exposure, the other displaying elapsed time.

Circuitry

The electronic control system comprises six basic circuits: control, time base, exposure measurement, aperture opening and closing correction, shutter drive, and power supplies. All circuits will be fabricated of modular printed circuit boards.

The control circuit will select the appropriate logical operations to accommodate the mode chosen by the user. This circuit will be constructed by using transistor–transistor logic (TTL).

The time base circuit will use a temperature-stabilized 100 kilohertz (kHz) quartz crystal in a self-starting Colpitts oscillator configuration. TTL decade counters will provide clock pulse trains at 1 kHz to 0.1 Hz. These pulses are counted to display the elapsed time and are digitally compared with the preset time when the timer mode is operating.

The exposure measurement circuit will use a chopper-stabilized operational amplifier with a low-leakage capacitor to integrate the optical power as measured by a silicon photodetector. The integrator output will be converted to a pulse train by using a precision voltage-to-frequency converter. Then these pulses are counted with TTL decade counters, and the digital readout displays the result. An analog comparator, using a voltage divider circuit, compares the integrated voltage with the preset voltage selected.

The aperture correction circuit will compensate for the finite opening and closing times of the shutter. Premeasured shutter characteristics will be used to set correction factors, and these settings will be manually changeable according to the shutter being used.

The shutter drive circuit will provide step voltages for the shutter solenoids. The first shutter (normally closed) is driven open, while the second shutter (normally open) is driven closed.

The power supplies will provide $\pm$ 15 volts for the analog circuits, a regulated +5 volts for the TTL circuits, and an adjustable 0 to +5 volts for operating the light-emitting diode (LED) digital displays.

The goal of this work is an instrument that achieves the high-accuracy specifications cited and an instrument that uses only standard, readily-available components in its construction.

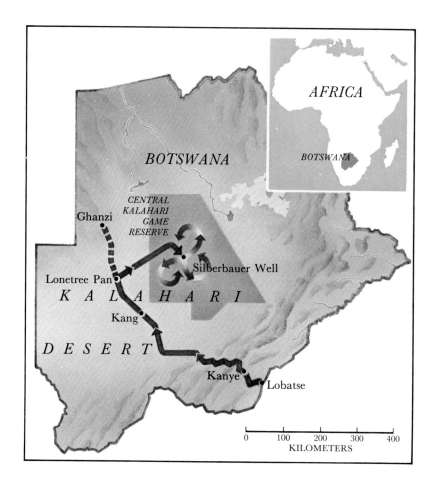

RECORDING A CULTURE DOOMED TO DISAPPEAR: KALAHARI BUSHMEN

It is difficult, at best, to face the prospect of losing one of our oldest cultures to "progress," even given contemporary notions that gradually it will be replaced with a better life for those involved. Can we really be sure that the creature comforts and improved quality of life our so-called civilized world can provide is truly a better alternative for a people who have enjoyed a long, rich cultural tradition of their own? Yet the argument will ultimately go against the primitive cultures, who will gradually—or sometimes rapidly—disappear by assimilation into an increasingly homogenized world culture.

As cultures disappear, we lose basic elements of our heritage and narrow our view of humanity. This project is designed to record a quickly vanishing culture, before its value for us is lost.

CARLOS PEDRO NOLASCO VALIENTE NOAILLES
Av. Figueroa Alcorta 3085
(1425) Buenos Aires
Republic of Argentina
Born January 31, 1932. Argentine.

A legal counselor, Carlos Valiente Noailles is professor of constitutional law in the law faculties of both the National University of Buenos Aires and the Argentine Catholic University and also is the author of books and publications on constitutional law. As background for this work, he took his law degree in 1955 as well as his Ph.D. in law and social sciences in 1966 from the National University of Buenos Aires.

In a quite different area, Carlos Valiente Noailles is an expert in ethnographic studies of peoples and tribes located in southern Africa and has written a book and several articles on them. He is a correspondent member of the Lisbon Geographic Society and a member of the Société des Amis du Musée de L'Homme of Paris.

I propose an expedition to the Central Kalahari Desert in Botswana to record songs, poems, legends, myths, folklore, and oral literature in general, of the Bushmen inhabiting that area. These Bushmen have reached a stage of material development comparable to that of Europeans toward the close of the Stone Age. Investigators have carefully studied the customs and characteristics of these people, and plenty of information of this type is available as a result of their efforts. But we still need to study and record their songs, poems, and oral literature. Time is working against us, as the Bushmen decrease in numbers and as many of them contact more advanced ethnic groups. A proper study of this almost unknown expression of art may soon become impossible.

Purposes of the Project

1. To obtain valuable cultural material that is doomed to disappear.

2. To add to our limited knowledge of the oral literature of so-called primitive hunters and gatherers. The project represents the only means of studying the origins of verbal art and literature of the human species in general.

3. To disseminate knowledge about an admirable people, praiseworthy for their will to survive and adapt to a hostile environment and for their artistic life (formerly rock and cave paintings; today, music and oral literature). Their achievement represents a triumph of spirit over the harsh conditions in which they live.

4. To help show the real value of these people to their countrymen.

5. To contribute to the spread of knowledge about an African culture that is today completely inaccessible to the Spanish-speaking world and only slightly known in the rest of the world.

6. To publicize the harsh living conditions of the Kalahari through the diary of the expedition, which should be of considerable interest to the general public. The expedition is a great challenge to document and record material in a hostile environment.

A hundred years ago, W. H. Bleek published tales, poems, and legends obtained from Bushmen. His sister-in-law, L. C. Lloyd, continued his work, and both efforts were published between 1875 and 1911. Their work represents the finest collection of Bushman literature to be published up to this time. It is also possibly the best collection of any primitive people's literature. These publications point out the sharp contrast between the outstanding beauty of such art and the astonishingly primitive stage of development of these people.

Dorothea Bleek, W. H. Bleek's daughter, also carried on her father's work, but neither she nor other investigators who worked in Bechuanaland (Botswana), Southwest Africa (Namibia), and Angola achieved any notable success with regard to Bushman literature (present-day authors of Bushman studies only make use of the original reference works).

My project is to organize an expedition to the Central Kalahari, the purpose of which will be strictly limited to collecting Bushman songs and oral literature. In my opinion, one of the main causes of failure to obtain suitable new material has been the multiplicity of objectives set by virtually all such ethnographic expeditions. They have tried to take a close look at practically every aspect of the lives of these people and have soon found out that they have bitten off more than they can chew.

This type of expedition calls for experience and capacity in overcoming innumerable difficulties, as the Bushmen live in remote and uninhabited areas. Day-to-day living is plagued with an unending series of problems, and the Bushmen tend to be chary of greeting strangers whose motives they doubt. As a result, when contact is made with a group of Bushmen, much ethnographic material receives only a superficial look.

My expedition would consist of one driver or mechanic, an assistant driver, a cook, a photographer (all experienced specialists who will relieve me of many worries and troubles), someone to make recordings, a Bushman guide, and three African interpreters familiar with the regional Bushmen dialects—mainly, G/wi, Naron, Hie, G//ana, and /aba.

The expedition would enter Botswana at Lobatse and follow the Lobatse–Ghanzi Road as far as Lonetree Pan, where a fresh supply of water can be taken on. The expedition would set out with 650 liters of petrol and 450 liters of water. The Bushman guide and one interpreter would join us at this point. We would then turn into the Central Kalahari, following the bed of an old river to the Okwa Valley (a two-day journey), which we would follow until we reached Silberbauer Well. We would camp here because there is plenty of water. We know the type of pump used at the well and would take the necessary spare parts to fix it if it is out of order. Supplies, including food, would be brought in every 12 days from Ghanzi by lorry.

We would start working with the Bushmen living around Silberbauer Well, with whom I became friendly during my recent expedition. As and when we decided the material had been completely recorded or that it was not what we were looking for, we would make sorties to contact other groups, taking with us one or two of the Silberbauer Well Bushmen to "introduce" us. These men would also act as guides, as they know where their fellow Bushmen live. (The Bushmen live in groups of 40–60 people, and each group keeps to a certain territory, which it never leaves, except to find food or to visit other groups.) We would stay wherever we could establish dialogue.

According to the established custom, the Bushmen would turn up at our camp at mealtimes to look us over and accept food and tobacco. We would feed them, gather round the fire with them, listen to their music and their talk, and play some of our music for them. Once this wordless form of communication—and that unique atmosphere of brotherhood and friendship born from being in company at nighttime in the lonely desert—had been established between us, dialogue would be opened through the interpreters. This speech would be discreetly recorded. Initially, questions would not be put to the Bushmen. The desired subjects would arise spontaneously, and we are fully prepared to wait for hours, or even days, for this to occur. Only when the atmosphere of friendliness had reached the stage where everyone was happy and ready to join in freely and naturally would we try to lead them by indications or questions toward the material we seek. The whole thing would be recorded carefully so that the translations by our interpreters could be properly checked, if necessary. Translations into English and Spanish would be made on the spot, with the help of the three interpreters, who, as far as possible, would be "cross-examined" separately, to check on their interpretations of the material.

The material obtained would be written and published in English and Spanish, along with photographs, an introduction to the Bushman culture, and an account of the expedition. A record of the music obtained would accompany the publication.

Varanus komodoensis, *two Komodo dragons.*

THE KOMODO DRAGON EXPEDITION

Komodo dragons do exist, you know, although our Western world had never seen them alive until 1926, and they do justice to all the portraits of mythical beasts you've ever encountered. Known to have reached lengths of at least 12 feet, with reports of much bigger beasts, they were able to go on with their long lives in virtual peace until well into this century. They could do so mainly because of their incredibly remote and circumscribed habitat, tucked away far from the beaten track in a handful of tiny islands.

They are still difficult to reach, but they no longer enjoy the solitude that was their greatest protection against people. The seemingly inexorable spread of the human animal is threatening what was once a balanced "dinosaur world" and endangering the dragons themselves. Too little of them is known, and lengthy scientific observation throughout their seasonal patterns is needed. To that end, this proposed expedition would capture on film these remarkable animals in an attempt to ensure their protection.

BARRY JOHN BROOKE
Rothsay House
1 Trevelyan Road
West Bridgford
Nottingham NG2 5GY
England
Born October 2, 1946. British.

After studying zoology at Nottingham University, where he gained an upper second class honors degree (B.Sc.) in 1967, Barry J. Brooke became interested in filmmaking. After having made a number of films on his own, he spent five months in Greece, where he filmed praying mantises and small lizards and learned snorkeling and marine biology. Now working on an educational film in marine zoology for use by scuba divers, as a project for the British Sub-Aqua Club Underwater Conservation Year (BSAC UCY), he finds time to take part in general biological surveys as part of BSAC UCY, in addition to his own business as a manufacturer, wholesaler, and retailer of clothing.

The general aim of my expedition is to visit the islands of Komodo, Pada, Rinja, the west coast of Flores, and other, smaller islets of this area of Indonesia. The expedition will report on, observe, photograph, and film the Komodo dragon (*Varanus komodoensis*) in its natural environment. There are two specific aims: to record on film the life history of the animal and to examine the relationship the animal has with the sea by obtaining, for the first time, photographs and films of the animal swimming at the surface and in the depths of the sea.

Attention will be given generally to observing and recording by photographs and movie film the behavior of the animal outlined in my general account, any behavior of the animal not previously recorded, and any variation in appearance and general behavior among the Komodo dragon populations of the several islands.

The Komodo dragon will readily take to the sea and crosses wide stretches (1300 feet) of water to obtain food. This aspect of the animal's behavior merits further attention, and any new evidence might help to confirm one of two hypotheses: that the animal actively distributed itself in the past by swimming from island to island from its source in Australia or that it was distributed by passively hanging on to floating vegetation. I hope that tagging the animals on each island will show any subsequent distribution.

The Komodo dragon is a threatened species. Very soon, the inevitable march of tourism and industrial progress will reach this hitherto sheltered area and disrupt the delicate balance between these large, rare animals and their environment. It would be worthwhile to record the situation before any change takes place and to use this information to promote conservation of these and all other animals.

A final objective will be to keep costs down and to redistribute any money awarded. After the expedition, I will attempt to set up a fund from some of the monies obtained for use of material gathered on the expedition. This money could be given to other conservation expeditions, thus perpetuating the cycle.

Several visits to the area will be necessary in 1977–78. In an initial visit, we will assess the situation and prepare for subsequent ones. I hope to make contacts locally (Bali and Flores) to obtain a boat, diving equipment, and some experienced divers to assist—preferably ones who know the area.

The expedition from England will consist of two people. We will fly from England to Denpasar in Bali and then travel the 200 miles or so to Komodo by whatever transport is available there at the time. The return journey will be by the same means as the outward journey. The method of travel and the duration of subsequent visits will be determined after the first visit. It is predictable, however, that one visit of six months or so will be necessary to film the life history of the dragons.

To illustrate the possibilities of the material to be obtained by the expedition, a general account of the behavior of the animal is presented.

General Account of the Life and Behavior of the Komodo Dragon

Classification: Class Reptilia
                  Order Squamata
                  Family Varanidae
                  Species *Varanus komodoensis,* Komodo dragon

The Komodo dragon is the largest living monitor lizard. The largest living specimen on record is 9 feet, 11 inches long. Skins of 12-foot specimens were reported in the 1920s, and before that stories were told of even larger 30-foot monsters. Today the males often reach the size of 9 feet long, some 10 feet, whereas the females are usually a more modest 6–7 feet in length. The young are born with a length of 16–20 inches.

The Komodo dragon lives only in its natural habitat in Indonesia and is isolated to a small area of approximately 50 square miles on the islands of Komodo, Padar, Rinja, the west coast of Flores, and a few of the neighboring islets. The population of about 6000 animals (World Wildlife Fund estimate) appears to be surviving well, but closer examination reveals that possibly only 400 mature females live over the entire range; this is perhaps the most critical factor in the survival of the species.

Furthermore, probably any substantial increase in tourism or the development of mineral and oil interests would soon lead to a decrease in numbers. Sooner or later, one or both of these developments will occur. The animal is therefore a threatened species. The Indonesian government has created reserves in southern Rinja and in the uninhabited Padar, has generally discouraged hunting and killing, and has forbidden export of the species. However, these restrictions have proved difficult to enforce on Komodo, where the local tribes hunt the dragons for food and regularly burn the vegetation, killing both the predator and its prey.

The Komodo dragon is a diurnal animal, spending most of the day between 8 A.M. and 6 P.M. foraging for food. In the early morning, it warms itself in the sun. At midday, it shelters from the extreme heat of the sun. The rest of the day is spent hunting for food, lying in wait for prey, catching and eating the prey animals, recovering from the exertions of the hunt, visiting its water hole, and finally returning to its burrow for the night. The lizard eats a wide variety of prey: grasshoppers, rats, birds, fish, goats, hogs, deer, monkeys, and occasionally even horses and water buffalos. It also eats bird eggs, turtle eggs, and the eggs of its own species. Generally, the larger the lizard, the larger the prey it will take.

The Komodo dragon is a reptile and depends on external sources of heat to maintain an optimal body temperature. Animals that do this are called *ectothermic*. Mammals, which are considered to be more successful than reptiles in their independence from the environment, are *endothermic*, maintaining their own body temperatures from within. An ectothermic animal has the daily problems of first warming up enough to hunt for its prey efficiently and in competition with other animals and of, second, not overheating itself in the exertions of the hunt. It has, in fact, a limited activity range. It has been found, however, that the Komodo dragons, possibly because of their large size, can maintain a reasonably constant body temperature with a fair degree of independence from the environmental temperature. This enables it to be an active and effective predator on the mammals of the islands. It is a hunter and a scavenger, and these two habits appear to be related. The Komodo dragon will lie in wait on a game trail and lunge at any animal coming within 3 feet. It usually clamps its jaws on the prey and hangs on. The victim struggles frantically to free itself. The dragon has sharp, backward-curved teeth that cut deeply into the muscles and blood vessels of the captured animal. It wedges its powerful legs against any struggles, occasionally jerking backward forcefully. Once the prey falls over, it is quickly eviscerated and eaten. If, however, the prey manages to escape, death still awaits it. The bite of the Dragon transmits bacteria into the wound and infects it. The prey may die of infection or may be weakened and thus more easily caught by another lizard. If it dies of infection, it will form carrion for a scavenging lizard. Either way, the dragon population benefits. There are, effectively, no other predators to take advantage of the system, although some feral dogs do compete for the carrion.

Once the prey has been taken, the dragons eat very quickly, extending their stomachs enormously. If surprised during eating, they have been known to regurgitate the whole meal, possibly to escape without the encumbrance of a large, dragging stomach.

The Komodo dragon has been described variously as a poor to moderate swimmer. It is certain, however, that it will take to water and to the sea. The dragon has been observed entering water and either swimming at the surface or sinking to the bottom. If chased while on the surface, it will play dead with its head under water and its forelegs over its back—a curious, unexplained behavior pattern.

W. Auffenberg noted an interesting situation on one of the islets, called Nusu Mbarapu, off Komodo. When the goat population on the islet is high, and when the prey population is low on Komodo, large monitors will swim 1300 feet of swift tidal currents to reach the islet and gorge themselves on the goats.

The Komodo dragon society is hierarchically based on size, with the largest dominating. They hunt alone, but their burrows are often close together, suggesting a possible social function. They gather mainly at carrion sites, to which they are attracted by the smell, which we find repulsive. There is much interaction here, such as courtship and breeding activities. Here the social hierarchies are established or reinforced. Here, too, the smaller lizards risk being eaten by the larger ones, which are not averse to cannibalism. If a large lizard is having difficulties in breaking up a large prey animal, smaller lizards will approach closer than is normal to tear off pieces of meat and so cooperate in separating the food.

Mating and egg laying take place between July and October, and the incubation period of the eggs is possibly as long as six months. The female lays a large and varying number of eggs in a hole in the ground that is soon covered over. No cases of care for the young have been noted. The males will eat the eggs or the young if they find them.

History

It is surprising that such a large animal as the Komodo dragon was unknown to Western science until as recently as 1926. P. A. Ouwens first described the animals from accounts and skins while working at the Buitenzorg Botanical Gardens in Java. He called it *Varanus komodoensis* Ouwens. In 1926, Douglas Burden, accompanied by his wife and E. R. Dunn, an authority on reptiles, took an expedition to Komodo and observed the reptiles for the first time. Typically, they shot the largest specimens they could find.

Subsequently, a great amount of interest in the animals was shown. A live specimen was exhibited in London in 1928, and various papers were published on the giants' morphology, behavior, and affinity with other species. Some speculated on how the lizards reached their isolated home because no dry ground route was ever in that area.

In 1936, Lady Broughton conducted "A Modern Dragon Hunt on Komodo." After that, interest subsided in the face of World War II. In 1958, A. Hoogerwerf visited Komodo, closely followed by the Russians in 1962 and the Americans in 1968. F. W. King was sent by the New York Zoological Society to report on the dragon and on the need for its conservation. This trip was followed by W. Auffenberg's one-year stay on the island in 1969–70. So far, this is the last scientific survey I can trace, and it appears to have been fairly comprehensive. I have written to Auffenberg for further details.

The Island of Komodo

Komodo Island is 22 miles long north to south, and 12 miles wide east to west. It is volcanic in origin, and the central mountains rise to 2400 feet. The valleys are covered in alang alang grass and low bushes. The occasional tall, spindlelike lontar palm trees hold their tufts of leaves tight and well out of reach. The rains fall only between January and March. Thereafter Komodo is dry and arid, choked by the powdery layer of gray volcanic ash. There is only one permanent well, in the small village of Komodo on the west shore at Telok Sawa (Python Bay). By August, the island has lost all trace of greenery and bakes in midday surface temperatures of 167°F. Life is renewed again in the rains of January, when streams flow from the mountains. The surrounding sea is not deeper than 200 feet. Sharks and sea snakes and an occasional giant lizard discourage all but the intrepid explorer with a camera.

A WAY TO STOP THE ANNUAL SLAUGHTER OF 200,000 PORPOISES

As we become more technologically adept at ensnaring great schools of commercially valuable fish, we have wreaked havoc on the ecological environment. Perhaps the most chilling effect of our skill is the killing of an estimated 200,000 porpoises annually as the by-product of tuna fishing. Despite efforts by conservationists, governments, and the tuna industry, the killing goes on, creating tensions both inside and across national boundaries. The problem is clearly defined in the project described here, along with a provocative potential solution that involves the help of one of the porpoise population's oldest enemies, the killer whale.

DANIEL A. SHEPARD
Honorable Mention, Rolex Awards for Enterprise
67 Kirkland Street
Cambridge, Massachusetts 02139
United States of America
Born March 17, 1921. American.

Daniel Shepard received his B.Sci. in psychology from Harvard University in 1943. Today, as a contract consultant with Arthur D. Little, Inc., a leading international research and consulting firm, Daniel Shepard is involved with interpreting the diverse technical and professional activities of ADL to its clients and the public. His principal fields within the company are the life sciences, bioenvironmental studies, information sciences, electronics, management counseling, telecommunications, and special projects, such as the emerging technology of fiber optics and the development of alternative energy sources.

I propose an electronic warning system for the commercial fishing industry, designed to exclude marine mammals, specifically porpoises, from the tuna catch during purse seining.

Porpoises and yellowfin tuna travel together, for reasons that have not been fully explained. The industry's shift from line fishing to purse, or lampara, nets that enclose fish from the bottom as well as from the sides, has resulted in the deaths of more than 200,000 porpoises (of six species) annually. The United States has therefore imposed restrictions that are not only difficult to enforce but also economically limiting. The US claim of 200-mile offshore fishing boundaries has further aggravated the situation.

United States tuna fishermen are threatening to reregister their boats under the flags of nations that do not have, or do not enforce, regulations against the killing of porpoises incidental to purse seining. This approach will continue to reduce the porpoise population to a point where unforeseen and disastrous consequences for the ecological cycles of the seas may ensue. Yet the electronic warning system would reduce, if not virtually eliminate, the danger to porpoises, most numerous of sea mammals, without affecting the tuna catch.

Background

Netting porpoises in purse seines (a method that became widespread in 1960) causes death by drowning or injury to the mammals. More than 3 million have died in this manner since 1960, according to the *National Wildlife Newsletter*. The situation is complicated by:

1. The well-known habit of tuna to follow porpoises, for reasons not clearly understood but possibly related to the superior navigational abilities of porpoises.

2. Failure of the porpoises' echolocational capabilities to detect the netting of purse seines. Their snouts become entangled in the nets, and they suffocate from lack of oxygen.

3. Epimeletic behavior of porpoises who, as members of the whale family, exhibit the social tendency to stand by and assist their kind in distress. Therefore, unnetted porpoises often refuse to leave those who are entangled, and thus they are drawn in with the rest of the catch.

4. Although tuna fishermen have adopted the practice of reversing their tuna boats just before hauling in, to free porpoises from the bunt (lower portion) of the purse seine, this has not proven wholly effective. And, although they manually disentangle the porpoises from the net and release them from the rest of the catch, the mishandling and shock often prove fatal.

The US Maritime Protection Act of 1972 sets a limit on the depletion of porpoises by such incidental killing and decrees a moratorium on tuna fishing once this limit has been reached. The reaction, which is a matter of record, is one of resistance by US flag fishermen. These regulations, moreover, are not binding on fishermen of other nations in their own or international waters.

The US tuna industry has claimed that it is close to finding methods that would virtually eliminate porpoise deaths. But they have not as yet revealed the nature of these methods or how they might be applied internationally.

Proposed Project

I suggest an underwater sonar technique to transmit (by tape loop) the sounds made by *Orcinus orca* (the killer whale), a relative of the porpoise but its enemy and a predator of almost every swimming creature in the seas. The device can be mounted on fishing boats, with smaller models mounted on the frames of the purse seines themselves to reinforce the warning. In the drawing, X's mark typical placements of the electronic warning units on the boom of the purse seine net and on the tuna boat hull.

The practicability of such a device is strongly suggested by the sophisticated fish-finding units that commercial tuna fishermen already use, both hull- and net-mounted, to search out large schools of tuna electronically.

An odontocete like the porpoise, *Orcinus orca* also echolocates through bursts of high-intensity directional sound up to 150–200 kilohertz. In the range of human perception (20 hertz to 20 kilohertz), killer whales also emit characteristic barks, screams, moans, and whistles, primarily for social communications. We propose to capture these forms of echolocation and social sounds of killer whales for retransmission by a directional device that will warn porpoises away from the open end of purse seines, without diverting the accompanying schools of tuna.

Research and Application. Sufficient numbers of killer whales are in captivity to allow recording of their sounds in and beyond the human-audible range and in the most effective combination for our purposes. The technology for transmission already exists.

There must be experimentation to determine the optimal distances from the nets to warn off porpoises before the less-perceptive tuna change course to follow. This will keep the mass of fish in motion toward the waiting nets. Again, both privately endowed and governmental marine institutes are capable of testing the feasibility of the method.

Enforcement. If the proven technology and technique were to be adopted as standard by an international organization, preferably UNESCO, and incorporated into the law of the sea, the means could be found for usage across national lines of interest. Moreover, progress toward common agreement on this point might provide an opening wedge toward agreement on matters of even weightier nature, such as exploration and exploitation of the sea bottom.

Nothing hinders an immediate start of the research and development activities that would substantiate the feasibility of the proposed project or hinders its early application from a technical standpoint.

LOW-COST ZINC OXIDE FROM RECYCLING WASTES

As a key chemical in a variety of industrial applications, zinc oxide presents a problem for many developing economies. Either it must be imported, or a sizable investment must be made in plant facilities required to produce the raw material.

In both cases, the costs of obtaining zinc oxide can be viewed as disproportionately high when capital is limited. To tackle this problem, an enterprising Indian engineer has developed a process that promises an alternative way to supply this needed ingredient to many small-scale industries. If successful, its application would be of value in many parts of the world.

UMA PRASAD MAHAPATRA
10/1 Geeta Society
Thana — 400 601
India
Born December 29, 1934. Indian.

As a graduate in science from Utkal University in 1955 and in metallurgical engineering from the Indian Institute of Technology in 1960, Uma Mahapatra is a consulting engineer and proprietor of Inventa Technical Services, which offers services to foundries and such industries as metal powders and chemicals.

He is interested in new ideas and inventions and has Indian patents on a new type of ingot mold and a dimensionally variable mold box. As a partner in Jaganath Chemical and Pigments, he is involved in a pilot plant for making zinc oxide from zinc wastes with a new process he has developed.

Zinc oxide, an important chemical in paints, rubber, ceramics, and other industrial applications, is generally made in two different ways: the direct, or American, process and the indirect, or French, process. In the first method, zinc ores are reduced and then oxidized to zinc oxide, and in the second method pure zinc is burned to form zinc oxide. Both processes require costly machinery and investments, but neither can control burning to make different grades of zinc oxide in the same unit simultaneously.

My project is based on a new idea of making zinc oxide from all types of zinc wastes—such as zinc ash, zinc dross, and zinc hydroxide—and also from zinc metal and/or mixtures in various proportions. With controls on burning rates, charging rates, entry of air, and temperature, this method, a golden mean between the two existing processes, can produce zinc oxide at low cost, without heavy investment in plant and machinery. Other features are the use of the same furnace in first reducing and then oxidizing, and the use of catalysts and fluxes.

Concept

The project has been conceived mainly because it is difficult to interest the public in an invention that has not been demonstrated to be successful. The inventor may have to do this first. Then, if no agency supports the project, it will have to be offered to industry. In the event of success, the project may generate enough capital to finance further research.

I have already proved the feasibility of the process on a pilot-scale operation with an output of about 500 kilograms per day. The pilot plant is shown in the photograph. This operation in itself is suitable for commercial operation because of the low cost of production. The next stage of the project will produce 2 tons of zinc oxide per day. As described earlier, this plant will use all types of zinc waste, such as ash, dross, hydroxide, granules, and die-casting scrap.

Description

First the raw materials are treated. Zinc-containing materials, such as galvanizer ash (containing zinc and ammonium chlorides) are pulverized, screened, and heated to 800° Celsius to eliminate volatile chlorides, cadmium, and other elements. Sulfide ores are heated to eliminate sulfur. Then the mix is pelletized by adding lime and a catalyst like lithium oxide to hasten the reduction of zinc. The flux draws out iron, aluminum, copper, and other high-boiling-point metals that must be removed.

The furnace is a vertical stack furnace—a miniblast furnace, fitted with a rotary floor and supplied with hot air. A bed of coke is made and heated. To this furnace, zinc waste pellets are periodically added. The required coke is added from the charging door. In the lower part of the furnace, zinc waste is reduced by carbon monoxide to zinc vapor. The zinc vapor, carbon monoxide, nitrogen, and carbon dioxide rise to the top. Suction from two blowers in a hooded cyclone provides a continuous blast of heated air, which oxidizes the zinc and carbon monoxide. The heated gas carrying zinc oxide and other vapors enters the cyclone, where the coarser particles settle. The zinc oxide is then collected. The zinc oxide is graded and treated separately as required for commercial use.

The amount of air blown in to the furnace at the charging door is controlled to regulate the rate of reduction and oxidation. Valves for entry at the cyclone and for entry of secondary air control the grain size of zinc oxide. The capacity of the suction blowers is periodically varied by dampers and pulley ratios.

The flexibility of the process lies in the fact that along with zinc oxide waste, metallic type or pure metal zinc wastes can be used in various proportions to regulate quality. Only the zinc vapors and not the metallic additions are oxidized; there is no prior reduction. Also, the process can be run continuously.

The grain size of the zinc oxide produced is finer, owing to turbulence while burning; used in paints, the fine grains offer good covering power and oil absorption. If heated longer, the zinc oxide grains grow in size; this can be achieved by controls either in the system or outside.

The impurities, such as iron, copper, lead, and aluminum, are slagged out by various fluxes and removed from the bottom of the furnace, mostly in dry powder form.

To produce leaded zinc oxide, lead can be added to the zinc, thus increasing oxidization. Here the height of the coke bed is the deciding factor.

The process needs about eight persons for three shifts. The controls are rigorous, and explosions may occur if the charging door is kept closed too long. Otherwise, it is simple to operate. This process uses waste materials that would otherwise pollute the atmosphere. Moreover, it needs very little electrical energy. And since the scale of operation can range from a few kilograms to tons, it suits small-scale operations.

For my project, I plan a unit of 2 tons per day. About 75 percent of the investment in plant and machinery will be obtained from a government loan. I am confident that I will be able to repay the loan within six months' operations.

CHANGING A POPULATION'S DIET TO PREVENT MALNUTRITION

Malnutrition and starvation often are confused, usually tending to be considered problems arising out of similar circumstances. Although the two often go together, we need to remember that starvation is caused by lack of food, while malnutrition arises from eating the wrong food or from the lack of adequate quantities of certain necessary foods. Given what we know today of the nutritional needs of the human body at its different stages of growth, it is doubly tragic that malnutrition continues to occur in regions where, considering the available or potential food supply, no technical reasons for it should exist.

The problem frequently occurs in a population that has followed a traditional diet, based on local resources, for long periods of time. Typically, the poorest members of society change diet most slowly, regardless of how available needed foods are. This is often, and most painfully, seen in the very young, in infants and children dependent on the knowledge and resources of their parents to provide them with food. This project sets out to do nothing less than permanently change the dietary habits of a large population.

JORGE RAFAEL RESTANIO
Honorable Mention, Rolex Awards for Enterprise
Patricio Diez 1286
Reconquista (Santa Fé) 3560
Republic of Argentina
Born February 11, 1932. Argentine.

After receiving his physician's degree from the University of Córdoba, Argentina, in 1958, Jorge R. Restanio became a pediatrician in 1969. As a rural physician, he works in pediatrics at Reconquista Central Hospital, the only public hospital in this city of 35,000 inhabitants (the hospital also serves another 50,000 inhabitants from the surrounding area). As head of the pediatric department, he concentrates on health care for children, plus running the clinic and the sanitary education program.

Because of his work, Dr. Restanio is constantly exposed to sectors of the population with malnutrition caused by local sociocultural factors. As a pediatrician, he is dedicated to teaching better diet standards to young mothers and to detecting dietetic deficiencies in children and adults. As part of his efforts to combat this local lack of knowledge about diet requirements, he teaches the subject of nutrition through 14 radio broadcasts in three countries and through booklets designed to educate people. He is very much a man with a mission, and the project he describes here is an ambitious and important one, of interest to many other communities with similar problems and opportunities.

I propose to demonstrate the possibility of making a definitive change in the dietary habits of a population with malnutrition, by means of balanced, low-cost feeding. My plan includes inducement and promotion, a proof period, and publication of clinical, psychological, sociological, and economic results.

This experiment will be done in Reconquista (Santa Fé, Argentina) because of the following considerations:

1. There is a high level of infant malnutrition in the area, especially in the 6-month- to 6-year-old category. During this period, the needs for caloric protein must begin to be filled by sources other than the cow's milk provided by public health authorities. Because of the high price of milk and meat, new sources must be found.

2. Adequate, locally grown cereals (wheat, corn, soybean, sunflower, peanut, and sorghum) are readily available.

3. There are adequate local flour mills to handle the different grains.

4. Also available is a meat-processing plant, capable of utilizing the entire cow carcass (meat, bones, blood, and so on).

5. Nutritional technicians are locally available.

6. A social service school is present in the area.

7. There is only one free medical center, which facilitates the periodic checking of the project's work.

We want to supply a mixture of inexpensive cereals and legumes, with or without animal proteins. The mixture will supply adequate levels of proteins and amino acids. These products (cooking flours, biscuits, and vermicellis) will be given as complements, or supplements, for use in the daily diet.

To convince people that this diet is worthwhile, we will use the appropriate techniques of social communication through health educators, experts in nutrition, and local leaders. The research will be done on families with similar economic conditions, habits, culture, and family structure. Prior to the program, all participants will be weighed and measured and their nutrition condition registered. They will be asked a series of questions: Do they wish to change their dietary habits? What are they presently eating? Who is the "boss" in the family structure? What resources do they have, and are they used? What are their rites and taboos? What is the mechanism of their food purchasing? What are their economic resources? What are their cooking methods? What cooking and eating instruments do they have? What do they think is their principal food? What do they think is the least healthy food? Do they suffer from diarrhea? Have they ever thought they suffered from malnutrition? To reach the people, we will use radio programs, posters, booklets, and so on, written in appropriate language and based on the answers given to these questions.

We believe we must work in three different places: home, school, and clinic. In homes, we will have a chance to convince people of the worth of this feeding method and also to teach cooking methods that avoid routine and loss of interest. In schools, the food will be given to the children at mealtimes. At the clinic (Reconquista Central Hospital), the food will be given to babies under a year old.

We shall assess the project through clinical results (weight, height, nutrition, digestion, and so on), psychological and social changes, the possibility of continuing the diet, and economic results (impact on family resources).

There is a great need to solve the hunger problem, especially for young children. My long years of work in pediatrics convince me that this project will be a new departure in applied science. I know of no other project that has tried to make definitive changes in the feeding habits of a large population. In fact, until now only a few clinical groups in several countries have experimented on a clinical scale.

In contrast to the usual slowness of social changes, feeding habits of the middle classes have changed rapidly because of advertising and promotion. These means of communication can be important instruments in producing similar dietary changes in the poorer social classes. No one has explored the effects of a continuous feeding education program by using the most developed methods of human communication.

My goal is to demonstrate that it is possible to change the feeding habits of a large population sector, to improve clinical and nutritional balance, to diminish infant mortality caused by malnutrition, and to help to solve the problem of hunger in the world.

A SHOE THAT CAN HELP MILLIONS OF PEOPLE TO WALK AGAIN

Some 25 million people suffer from the once dread disease of leprosy, against which modern medicine is finally making strides. Nevertheless, when the disease has progressed to a certain point, loss of nerve tissues removes one of the body's most important warning systems—the ability to feel pain at points of injury.

As leprosy tends to be rural and tropical, the people who suffer most from it seldom wear adequately protective footwear. Damage to the feet is not felt, ulcers develop in neglected injuries, and the feet ultimately become incapable of maintaining the victim's balance.

For some eight years, this dedicated entrant has attempted to counter this problem through the enterprising use of modern technology, in a program adapted to the special needs of the countries wherein leprosy has its strongest hold and most devastating effects.

GEORGE WILLIAM HALL CLARKSON
Honorable Mention, Rolex Awards for Enterprise
173 London Road
Clacton on Sea, Essex
England
Born June 16, 1912. British.

As a partner in a small chain of retail footwear shops, George Clarkson's interest in the project he describes here is an understandable one. Trained in footwear construction, he spent four years in the medical service during World War II and immediately thereafter served as a United Nations relief officer inspecting refugee camps in Italy. He has spent over eight years working with his project, which was originally prompted by a request for help from a Tanzanian leprosarium. Since then he has become a sought-after expert on the subject and is devoted to seeing the work of the project carried on.

People suffering from leprosy lose all feeling in their feet because the disease kills the nerve tissue. When they damage their feet with thorns, sharp rocks, and so on, they do not feel any pain and so do nothing about it. Ulcers resulting from this neglect in time affect the bones, causing osteomyelitis. Gradually the foot distorts, and the patient loses balance and no longer can walk.

My object in this project is to make special footwear that can restore a patient's balance; I do this by using materials that will stop further ulceration and possibly contribute to the healing of the ulcers. Leprosy is a disease of the villages, mostly in remote areas, and it will be necessary to train local people to make this special footwear and keep it in repair; therefore, it must be a simple method that requires only hand tools.

To start this project, I visited a leprosarium in Tanzania. There I witnessed many patients lying in hospitals because of large foot ulcers. The only cure appeared to be bed rest, which could take up to one year. Even afterward, there was a danger of their feet breaking down once they began walking again.

I visited the Institute of Orthopedics in London, where they have started to use Plastazote to make footwear for arthritics, who likewise have distorted feet but no ulcer problems. The conditions of these people and the leprosy sufferers are quite different, but the visit helped me form ideas for leprosy patients. The big difference is that the London workshop has lots of machinery, whereas in the remote bush only hand tools are available. Also, the bush people must continue working in spite of their affliction, because there are no disability relief programs to help them. As the problem is a tropical one, the shoe should not be completely enclosed, to ensure that air gets to the foot. Also, the patients frequently have no fingers; thus, special snap fasteners must be used for the straps so patients can fasten and undo their shoes. The shoes must be designed to restore the patient's balance. The shoes must be as light as possible, but the length of service they need to give must also be taken into consideration.

We must keep the project as simple as possible, because the people to be trained to carry on have little or no formal education. Hand tools are simpler to master, and machinery is expensive and difficult to set up in the bush. One necessity is a small portable electric oven in which to heat the Plastazote, which is then molded to the shape of the foot. If there is no electricity, a small generator is necessary. As timing to the second is critical for the Plastazote, a reliable timer is necessary.

When heated, Plastazote can be molded to the shape of the foot, and a shoe built round it, thus restoring the patient's balance. If these shoes are properly made, the patient can walk immediately, even if he or she has not walked for years. The patient can go straight back to work, which is important in countries that do not have, and cannot afford, relief payments for sickness. Moreover, Plastazote allows the ulcers to heal. Of course, many people have not yet reached the state of being unable to walk but are slowly losing their balance as their ulcers get worse. These people can be restored to normal walking.

Only within the last 100 years has anything been done for leprosy sufferers; previously they were shunned and feared. The Christian missions, both Protestant and Catholic, have taken an increasing part in caring for these unfortunate people. Now, many of the developing countries where this disease is rife (an estimated 25 million people suffer from leprosy) are allocating a small amount of their meager health funds toward this work. Most people suffering from leprosy are not receiving treatment owing to lack of money and workers.

For about 10 pounds sterling per year a man or woman can be kept in suitable footwear and so be able to walk and work. But this extra cost is more than most leprosy centers can afford without making sacrifices in other ways. This means fewer people will get treatment.

In Madagascar, Fame Pereo of Canada has allocated funds to

Catholic Relief Services to enable voluntary workers to tour the country instructing mission and government leprosy centers on how to make this special footwear. The success of the project has been outstanding, and the people who are able to walk again, or to see their friends walking once more, attribute it to miracles.

What has been done in Madagascar needs doing in all the developing countries. It was started in Tanzania, where a local young man goes round the country making these shoes, but he is hampered by the high cost of Plastazote. A scheme started in Togo seems to have collapsed for lack of funds.

Plastazote, special buckles, and strapping must be imported from England. The cost for a pair of shoes is about one pound, but prices keep rising. Moreover, freight is very expensive. Rubber for outer and inner soles, leather or plastic for covering the Plastazote, and glue are purchased locally; prices of these materials vary considerably from country to country. Overall, the cost of materials is about 3 pounds for a pair of shoes. Instructors from England and the United States have worked voluntarily, as have Christian missions, but local labor must be paid for, along with traveling expenses. The aim is to train local people to make the special shoes and take over, but they need financial aid.

There is no end in sight to the work that must be done. The biggest cost problem is the Plastazote and buckles, which have to be imported, as well as the portable oven. Funding from Europe or the United States would be of enormous help. The cost of all this may seem great, but is it, when measured against the fact that for about 10 pounds a year a man or woman might again walk and work?

IDENTIFYING THE CANCER HAZARDS IN OUR ENVIRONMENT

Behind the frightening statistics on the incidence of cancer in our industrialized societies, an increasing body of evidence suggests that the causes are environmental. If so, no matter what our eventual success rate in curing cancer becomes, we are well advised to start preventing the reasons for it in the first place.

Simple testing procedures designed to identify hazardous elements in our environment would be a boon to medical authorities the world over. This project suggests just such a procedure. As the entrant was accorded honorable mention by the selection committee, the project is noted here, although without full details, pending the outcome of a patent application. Interested readers can write directly to the entrant, who may be able to discuss the procedure by the time this book is published.

IMRE FEDORCSAK
Honorable Mention, Rolex Awards for Enterprise
Department 1
Institute of Biochemistry
Semmelweis University Medical School
Puskin u. 9
1088 Budapest
Hungary
Born June 19, 1926. Hungarian.

Following his 1954 graduation from Eötvös Loránd University's faculty of natural sciences in Budapest, Imre Fedorcsak went on in 1962 to take his Ph.D. in genetics, biochemistry, and microbiology at the same faculty. Later, he earned his candidate of sciences (1968) and his doctor of sciences (1974) from the Hungarian Academy of Sciences. As a professor employed by the Hungarian Academy of Sciences, he is appointed as leading research associate to the Institute of Biochemistry, Department 1, Semmelweis University Medical School in Budapest.

Dr. Fedorcsak's project, for which he received an honorable mention in the Rolex Awards for Enterprise, is entitled: "A New Biochemical Method for the Demonstration of Sister Chromatid Exchanges in Human Lymphocytes: A Routine Screening Test to Detect Environmental Mutagens and Carcinogens."

SOLVING CONSTRUCTION PROBLEMS IN LOW-TECHNOLOGY AREAS

There is a streak among the Rolex enterprisers that could be characterized as "Let's get on with it." They see a problem or an opportunity, and they get right into action.

The device described in this project has that flavor. It is practical, simple, elegant in concept, and appropriate. And it challenges those who wring their hands in despair over our inability to provide many parts of the world with adequate housing, plumbing and sanitation facilities, and the myriad other "built" items of reasonably acceptable community standard of living. This project says, "Here's a solution you can use today, yourself. Why not start now?"

OLIVER MATHESON BULLEY
5, Nomis Park
Congresbury
Avon BS19 5HB
England
Born August 9, 1936. British.

After completing the Royal School, in Wolverhampton, England, in 1955, Oliver M. Bulley went on to take his diploma at Loughborough College, before becoming a schoolmaster in the specialized area of technology and design.

My aim in this project is to develop and market, in conjunction with a hand-operated clay extruder, a range of extrusion die plates to produce building materials, drainage and irrigation pipes, and so forth, all designed specifically for low-technology environments. The particular features of this project are (1) the quick production of serviceable pipes and building materials, using a limited amount of labor and skill, and (2) the absence of any electrical or other motor power source, the scheme being designed and applied to a hand-operated tool. The initial proposed designs are for self-locating, low thermal conductivity bricks; dual-purpose, double-skin tiles; screen bricks; land drain pipes, and irrigation and sewer pipes. These items would be formed from local clay, using a hand-operated clay extruder.

The self-locating brick would have fired cross-section dimensions of approximately 80 mm × 115 mm and could be formed in any length up to 750 mm. The proposed design is based on the Italian low-thermal conductivity pattern, with the addition of self-locating ribs. In manufacture, the three primary advantages are (1) low material requirements; (2) when the bricks are stacked for drying air flows through them (in this way, tolerances are maintained with no special racking); and (3) no special slow firing is required (the lattice section of the brick allows firing at the same rate as in ordinary pottery ware). There are also three advantages in use: (1) bricks self-locate and lock when stacked, requiring minimal mortar binding, and, if necessary, can be used without mortar; (2) the bricks have low thermal conductivity; and (3) bricks can be used, if necessary, in the raw state.

The proposed tile design has a double-skin form with features similar to those of the bricks. Having a double skin, the tile can be closed off at one end in manufacture to exploit the low thermal conductivity effect, or it can be left open to provide a ventilating air flow. The flat tile may be used for roof or floor.

A screen brick pattern has also been included. This can be stacked and fired in the same manner as the bricks and tiles.

Standard 100-mm-bore land drain, sewer, and irrigation pipes, with collars, can be formed from similar die plates.

With a hand-operated clay extruder, approximately 50 kg of clay can be processed in five minutes.

SOME BATS ARE DEADLY, SOME ARE NOT — BUT HOW TO KNOW?

It is easy enough to determine that a disease is caused by a carrier host but difficult to ensure that the carrier host can be identified by those in a position to treat or prevent the disease. Without the resources of a laboratory or the expertise of authorities, identification can be frustratingly complex.

Disease-carrying bats are ubiquitous in South America. For the most part relatively harmless, they nevertheless pose a major health problem as transmitters of diseases dangerous to people. Although scientists and professionals in the field are generally well informed on the role of bats in human health, the general public has little, if any, information available on the problem. This project seeks to bridge that gap in the hope of saving lives.

REXFORD D. LORD
Apartado 2128
Las Delicias
Maracay (Aragua)
Venezuela
Born July 31, 1927. American.

As a scientific advisor on public health ecology for the Pan American Health Organization (WHO), Rexford D. Lord is involved in investigations and training with the rabies and arbovirus departments of the Instituto de Investigaciones Veterinarias in Maracay, Venezuela. His degrees, in zoology and vertebrate ecology, include a B.A. from Pennsylvania State University (1950), an M.S. from Texas A&M University (1953), and an Sc.D. from Johns Hopkins University (1956).

The purpose of my project is to provide a reference to help professionals other than chiropterologists identify South American bats. Workers in many fields are investigating the roles of bats in their specialties, and in South America they are encountering difficulties in species identification because appropriate references are scattered throughout the literature.

The role of bats in relation to the health, welfare, and economy of humans has been well documented. Confronted with this problem in my investigations of the ecology of rabies and arbovirus infections, I was forced to put together my own compilation, frequently on my own time and completely at my own expense.

From the beginning, I planned my project to incorporate many features of other useful guides to species identification, such as W. H. Burt and R. P. Grossheider's *A Field to the Mammals* (Houghton Mifflin, 1952). Thus, the guide includes a detailed description of each species, a map showing the distribution of each species in South America, and color photographs of each species, including a close-up of the face (facial features of living bats are useful distinguishing characteristics). The guide contains keys to the families, to the species within each family, and, with the large family Phyllostomatidae, to the subfamilies.

Each species description contains the scientific name; a common name (frequently coined); a physical description emphasizing special distinguishing characteristics but always including a description of the color, the nose, the length of the forearm (indicating size), and the weight (also indicating size); the dental formula; the habitat, including roosting habits; food habits; and relative abundance.

Workers using the guide will have in hand the specimen to identify. Following the keys, they will arrive at a tentative description. Then they will refer to the species description, distribution map, and photographs to determine if their identification is correct. No key is perfect, and beginners sometimes make some rather absurd identifications. Such gross errors are quickly revealed when one refers to detailed species descriptions and photographs. However, when species closely resemble other species, the differentiating characteristics will be delineated.

As we know, "A picture is worth a thousand words." The book *Bats in America*, by R. W. Barbour and W. H. Davis (University Press of Kentucky, 1969) illustrates the value of quality photographs. Yet most artists' renditions of bat species presently available are not good. The illustrations in my guide are of living bats brought into the laboratory and photographed in settings that match the animals' natural habitats. Because the bats are free, they often fly about the room and must be recaptured and placed on their perch again, which requires patience.

The guide was begun about a year ago and presently lists 193 species. I have insufficient information to complete the species description on about 10 percent of the species. This information is available in the libraries of the US National Museum and the American Museum of Natural History and will be obtained on my next "home leave." This information will also make possible completion of all the keys, which are presently about half finished. To date, I have photographed only 52 species. It is clearly impossible to photograph all species (some are known from only one specimen); therefore a goal of about 150 species has been set, to assure inclusion of most species that are likely to be captured by investigators. To achieve this goal, I will have to travel to regions of the continent where these species occur. My official work sometimes takes me to these regions, but I cannot plan such trips; they come sporadically and on the demand of others. My personal finances are not adequate to meet this need.

Finally, in addition to serving the needs of public health and animal health investigators, the guide should aid all ecologists interested in describing and preserving rapidly disappearing diverse ecosystems in South America. Many species are highly specialized, occurring only in restricted habitats, and as such serve as indicator (i.e., descriptive) species of specific ecological communities. The project started about January 1976 and may be completed by July 1978.

IS IT ATLANTIS, OR SOMETHING ALTOGETHER DIFFERENT?

Legends persist of civilizations long lost in the mists and upheavals of unknown times, and for many afford the welcome respite of dreams about more appealing, simpler times, free from the pressures of today. Not everyone, however, is content to dream. Bitten and bewitched by the world's riddles, these exploring souls will not be content until a skein has been unraveled, no matter what the odds may be.

Atlantis! Did it exist, as folklore would have it? If so, where? What does one do when a mysterious underwater phenomenon defies the world of modern science and beckons invitingly to those who would know its secret? There is such a challenge to the inquiring and the curious. Just off Bimini Island. There may be no man who knows more about it or who is more determined to wrest its history from the sea than this entrant, who will take you under the seas with him in search of a legend or a new discovery.

DAVID DANIEL ZINK
P.O. Box 653
Sea Pines Station
Virginia Beach, Virginia 23451
United States of America
Born September 17, 1927. American.

Currently professor of English at Lamar University in Beaumont, Texas, David D. Zink earned a bachelor of journalism (1952) from the University of Texas, an M.A. in English (1957) from the University of Colorado, and a Ph.D. in Victorian literature (1962), also from the University of Colorado. But those credentials do not tell the full story of a man who is an experienced mountaineer, blue-water sailor, scuba diver, underwater photographer, field electronics expert, surveyor, and jungle traveler. Fascinated with the riddle of Bimini Island's offshore puzzle, which he here describes, he has made six expeditions into the ocean depths of the area.

Near North Bimini Island in the Bahamas may ultimately be a major site of a heretofore unknown pre-Columbian culture. Estimates of the Atlantic sea level suggest that the last plausible human occupancy of the site was about 6000 B.C.

During the six expeditions that I led from 1974 to 1976, I completed the first full underwater survey of a megalithic site locally known as the Bimini Road. These surveys have revealed objectives whose successful pursuit will require several more substantially funded expeditions.

The underwater structure consists of stones of megalithic size (many weighing 15 tons) arranged in a reverse J shape about 600 meters long and 100 meters wide. Composed of limestone, these rectangular blocks are 75 to 100 centimeters thick, and, in the horizontal plane, typically 3 by 4 meters.

During these investigations, I found evidence that contradicts the reported results of an expedition sponsored by the National Geographic Society. For instance, three fractures found in the marine limestone of the seabed do not coincide with the joints of the megalithic blocks above. Analysis of the geological samples suggests a nonhomogeneous cementing pattern from block to block. For example, one block of shell-hash may reveal (in microscopic analysis of thin sections) an aragonite crystalline cement, whereas the adjacent block may be dominated by a sparry calcite cement. This evidence works against the claims of some critics who see the entire site as a homogeneous formation of beach rock. In other words, the cementing pattern so far perceived is not consistent with the idea of a homogeneous chemical environment more usually associated with natural formations.

But other types of evidence also support the archaeological hypothesis. During the 70 days of the 1975 expedition, two artifacts were found, neither of which were related to any known culture of the region.

In August 1976, on the fifth expedition, a survey of the buoyed site 800 meters offshore was made from the beach. This survey was based on sun azimuths employed to work around suspected magnetic anomalies. The results seem to confirm their existence. As reported in *Izvestia*, a Soviet engineer, A. I. Yelkin, recently has said that these magnetic anomalies in the area are related to lunar phases and has cited them as a possible cause in the Bermuda Triangle disappearances. (In Chapter Six of Martin Ebon's *The Riddle of the Bermuda Triangle*, I had previously suggested a possible connection between the disappearances and the remains of an ancient civilization.)

Aerial photographs clearly show the unusual orderliness of this megalithic site. One of several discontinuities between the site and its natural environment is the fact that the central axis extends obliquely across the ancient, drowned shorelines found in the vicinity.

Also, aerial photographs of sea floor patterns near the megalithic site (and at two other major locations around Bimini) suggest that an electronic profiler could pinpoint structures now buried in the sediments. The megalithic site rests on marine limestone, but sediments lie seaward and to the east and south of the island. Previous stone structures have been located in the vicinity by scanning the Turtle grass growing on the seabed. The patches of grass are evidently patterned by chemical discontinuities in the sediment. One aerial photograph indicates one such suspected underwater structure, which has since been silted over and will have to be relocated electronically.

To make most efficient and economical use of the profiler, present aerial photography of the sea floor in the vicinity of Bimini must first be upgraded in a photogrammetric survey. One of my colleagues has devised filtration methods to deemphasize the blue recorded by color emulsions, thus allowing substantial improvement in aerial photography of sea floor patterns.

The objectives of this proposal are confined to water no deeper than 10 meters. There is reason, however, to expect structures in much deeper water. Justifying the expense involved in exploring substantially deeper water will depend on the success of the present project.

In summary, the project involves these stages: photogrammetric aerial survey, profiler survey, and underwater excavation of suspected structures. The two surveys are planned for completion by the end of the summer of 1977, and the excavation is planned for 1978.

MAKING LABORATORY EQUIPMENT FOR SCHOOLS

If the tools of education, for one reason or another, are not available within a school system, gaping holes are left in planned curricula, and the building-block process of education is arrested. Modern education requires ever more equipment, particularly in the area of science.

When budgets limit the availability of much-needed laboratory equipment, science teaching suffers, *unless* someone tackles the problem with an inspired alternative, as in this project, which suggests that students make their own equipment and learn much about science in the process. The idea of youngsters making their own precision instruments in Sri Lanka ought to challenge school systems all over the world.

LEELARATNE SENANAYAKE
c/o Vidya Silpa
No. 19, Meetotamulla Road
Wellampitiya
Sri Lanka
Born November 8, 1927. Sinhalese.

At Ananda College, Colombo, in Sri Lanka, Leelaratne Senanayake earned his government senior school certificate, with exemption from the London matriculation, in 1945. As a technical assistant in the department of meteorology, Sri Lanka, he is self-educated in horology, and successfully completed the "B"-phase course in meteorology at the Meteorological Department Training School of Poona, India, in 1972.

B efore Sri Lanka's independence, and for a while after, all science equipment for schools was imported, generally from Britain. Education is provided free in Sri Lanka. Nearly 690,000 children, grades 6-9, study science in schools. This has necessitated the creation of nearly 6500 school laboratories. The government spends approximately 4,000,000 rupees annually to equip these laboratories.

With the rising cost of production in developed countries, the cost of equipment rose at an alarming rate. Thus less equipment could be obtained from allotted funds, and fewer schools could maintain good laboratories. These factors brought about overcrowding; often students were only able to watch experiments being done. Individual experimentation was almost dispensed with in many schools.

The government granted incentives for starting various industries, to open up employment opportunities as well as to conserve foreign exchange assets. Industrialists visited foreign lands, gathered technical know-how, and started many industries. But, seeing the sophisticated instruments, the extreme care required in the manufacture of laboratory equipment, and the limited, exacting market, they decided to invest in more profitable enterprises.

Meanwhile, the number of children increased, facilities decreased, and the pressure on the few schools equipped to teach science greatly increased.

In 1974, I, as a former science teacher who understood the situation, began this project. I found a goldsmith with years of experience and two meteorological assistants, one experienced in watch making and the other a researcher of production methods. Thus an enterprise that came to be known as "Vidya Silpa" was organized. The tools used were hand tools, and the raw materials came from scrap yards. The group spent hours and hours experimenting, creating, testing, and redesigning various laboratory equipment. Once the group was satisfied that their instruments would work, and would continue to work well in the ungentle hands of schoolchildren, the products were shown to the education ministry officials responsible for testing and purchasing.

Some skeptical officials believed that making one or two models was one thing, but that quantity production was another. This attitude may have been fostered by Sri Lanka's recent political past. In the pre-independence era, the children who could go to a school teaching science studied in the English language, using English texts and instruments. But after independence the medium of instruction was in Sinhalese, using the same instruments. At this time, foreign products achieved an unusual importance for the English-educated and an unwarranted insignificance for the Sinhalese-educated. Children who had to learn with foreign instruments—very often not handling them, but being made to watch demonstrations by teachers—became apathetic about science education.

The skeptics may have been of this group. Nevertheless, after subjecting the instruments to stringent tests, they ordered equipment for magnetic experiments.

More youths were recruited, machinery was built out of other discarded machines, and the orders were met. They included magnets of different shapes and sizes, magnetic plotting needles, compasses, deflection magnetometers, and dip circles. The only components that were imported were synthetic jewels from Switzerland, for moving instruments.

The project caught the imagination of the workers, who originally joined to earn a living but who stayed on to become participants in a national project, in spite of long hours, exacting standards, and lack of automatic machinery. They built their own jigs and fixtures to facilitate production and enjoyed creating instruments that they had been led to believe could only be made with highly sophisticated machinery. The enthusiasm, ingenuity, and skill of the youths in the project produced, in two years, instruments that, when shown at exhibitions, drew the inquiry, "Why are imported instruments exhibited as products of local industry?" Pride of participation proved to be an added incentive. Now the children are able to handle our own Sri Lankan equipment in the schools. They can feel that science also is an indigenous product, like their mother tongue. This attitude has also justified Vidya Silpa's low-price policy, to make equipment available to more schools.

The needs of the education ministry have been such that no permanent installations could be undertaken. Also, the needs of school laboratories varies, from magnetic instruments, to electrical, to heat-related, and so on. Thus most of the machinery has had to be built, dismantled, and then rebuilt for another type of operation.

Only one departure from well-known designs has been made so far—a chemical balance for use in school laboratories. Normally, a student takes from 5 to 15 minutes per weighing. Thus, in a class of about 50 students, for example, not all could do their own work. Vidya Silpa has introduced a balance that allows weighing in 2 to 5 minutes. Readings on this balance can be made direct up to 10 grams, to an accuracy of 0.1 gm. Capacity can be increased up to 300 grams. The cost per balance is 600 rupees, contrasted with 1500 rupees for an imported balance. The saving in time for an average-sized class is obvious, and the opportunities afforded to children to understand the fundamentals of using a balance are increased.

Quality control tests are made at the component stage, as well as at the assembled stage. Because of the diligent application of the workers in the project and the lectures that explain to them what a particular instrument is expected to do in a laboratory and what the workers are expected to do to keep the finished project functioning accurately for a reasonable length of time, only about 0.2 percent of the instruments have been rejected.

Since 1974, Vidya Silpa has grown from 2 to nearly 70 workers. The annual output is a total of 20,000 units of different types of instruments: standard resistances, post office boxes, magnetic plotting needles, gold leaf electroscopes, spring balances, calorimeters, triple-beam balances, switches, resistance boxes, magnets (different types), magnetic compasses, induction coils, magnetometers (deflection), dip circles, demonstration electric motors, two-way keys, and so on.

What of the future? To meet the needs of Sri Lanka school laboratories, production must be increased. We need to invest in production machinery. Foreign aid would enable Vidya Silpa to divert more funds to its experimental section, to develop a still wider range of instruments, which are much needed in a developing nation such as Sri Lanka.

ENTERPRISES IN BRIEF

Hartmut Walter, 3404 Butler Avenue, Los Angeles, California 90066, U.S.A.

PROJECT BAOBAB: A SYMBOL FOR CONSERVATION IN AFRICA

The oldest and largest surviving tree giants in Africa's savannah ecosystem are the baobabs (*Adansonia digitata*), well known by Africans and tourists alike and even revered by some. This project seeks to locate, identify, and officially protect the 100 largest and oldest of these fascinating life forms, to research them, and to promote their use as a pan-African symbol of conservation and environmental awareness.

Sister Margarita Revilla, St. Joseph Hospital, Santimen, Pingtung, Taiwan 900, Republic of China

A NONPROFIT HOSPITAL FOR SOUTH TAIWAN MOUNTAIN ABORIGINES

Fifteen years ago, it was difficult for this applicant to obtain permission to visit the mountain aborigines, whose customs once included headhunting. Through long efforts, she has introduced methods that have dramatically reduced infant mortality and disease, even though the mountain people must come to central locations for help. She would now like to build a local hospital to serve them better.

Douglas Crombie Anderson, "Rosebank," 13 High Street, Elie, Fife, KY9 1BY, Scotland, United Kingdom

THE SCOTTISH TIERRA DEL FUEGO EXPEDITION

This 8-man expedition, using their own, self-made, ferrocement boat, specially constructed for the exploration as an 18-meter mobile base, seeks to go through the Straits of Magellan to explore the mountain ranges in the north of Tierra del Fuego and on the western edge of the Patagonian ice cap, landing at points previously untouched by other expeditions, in order to document and map access routes, passes, and mountains.

Malcolm Anderson, 4 Lunsford Manor, Lunsford Cross, Bexhill on Sea, Sussex, England

THE SEARCH FOR THE ANCIENT SEA PEOPLE OF THE MEDITERRANEAN

Both archaeologists and historians have long been puzzled by the early references to a "Sea People," located in the Mediterranean and mentioned by the Egyptians (Rameses III's temple writings) and in the Bible (Genesis and Exodus). Preliminary research by this entrant indicates that a civilization located on Malta well prior to 1200 B.C. may have played a large part in spreading culture throughout Europe.

Nils Westby, 2857 Skreia, Norway

A NEW HANDLOOM FOR HOBBY OR INCOME

Traditional handweaving has given way, for the most part, to highly mechanized looms capable of intricate patterns not producible on previous small looms. This inventor has devised a breakthrough home loom—light, portable, and collapsible, which solves many of the prior limitations and places the rewards of handweaving within reach of those who would use it for personal pleasure or to earn cottage-industry income in Third World countries.

Roman Alvarez, Instituto de Geofísica, UNAM, Mexico 20 D.F., Mexico

ARID-LAND WATER TRANSPORT WITH SOLAR ENERGY

The development of a system consisting of a water reservoir (the evaporator) in which solar energy is used to produce water vapor, which is then suctioned off by a solar pipe, which can carry the water vapor over tens of kilometers to its destination, where it is condensed once again into usable liquid. Testing of the solar pipe indicates significant amounts of water can be transported in this fashion, particularly when aided by a pulling fan at the receiving end.

T. C. Chandran, Kaira Dist. Cooperative Milk Producers' Union Ltd., Anand 388001, Gujarat, India

A COMMUNITY-OWNED POWER AND FERTILIZER PLANT

This project involves the development of a community-owned gobar (biogas) plant, which provides not only badly needed gas fuel to the community but also an extremely rich fertilizer, in the form of the waste slurry produced by the plant. Conversion of India's production of some 400 million tons of wet cow dung per year to this form of energy and fertilizer would help countless communities to help themselves.

Geoffrey R. Woodford, Patches Farm, Buxton, Norwich, Norfolk, England

A WAY TO END OVERBOARD DROWNINGS

This inventor has created the Norwester Flotation Suit, a working coverall to be worn at sea, that becomes an emergency survival system. Possessing its own built-in buoyancy, utilizing 2-mm closed-cell foam, it becomes a thermal system, then an independent flotation system, and, in the case of slow rescue procedures, a total survival system. This suit is already in use and highly esteemed.

Karl Elis Bowin, Haroldsgatan 9, 216 17 Malmo, Sweden

ARRESTING THE TILT IN THE TOWER OF PISA

This proposal involves changing the foundation of the Tower of Pisa from a spread foundation to a *floating pile* foundation, by transferring the total weight of the tower to 20 concrete piles, connected by hydraulic pumps that will serve as adjustable bearings. This will make it possible to keep the load distributed as it is now, therefore always maintaining the Tower of Pisa in its present leaning position.

Alberto Bernardo Araoz, Juan B. Justo 1664, 1602 Florida, Buenos Aires, Republic of Argentina

FLYING THE ANDEAN WAVE—1700 KILOMETERS BY GLIDER

The project is to explore and utilize the wave phenomena that take place in the upper troposphere (6–12 km) in the lee of the Andes Mountains in western Argentina, between the 28th and 44th parallels. Using a glider as the main research tool to acquire various scientific knowledge, a number of long-distance flights will be attempted, including a 1700-km nonstop flight from Esquel to Catamarca.

Eyal Gilead, 6a Ehud Street, Haifa 34 551, Israel

THE ECOLOGY AND AQUACULTURE OF THE SPINY LOBSTER

Some ten years ago, there was a large population of the spiny lobster in the Bay of Eilat, with weights reaching more than 4 kg (8.8 lb). Their numbers and size have dwindled rapidly. This project involves the nighttime underwater study of the lobsters in their natural habitat, in order to determine population, distribution, and ecological criteria. It includes the laboratory testing of reconstructed environments for the captive breeding of this species, which is now used locally for food.

Maynard Re Mine Knisley, 7622 Kennesaw Drive, West Chester, Ohio 45069, U.S.A.

"REDUCE THE SIZE OF HUMAN BEINGS BY TWO-THIRDS . . ."

Admitting the first-glance bizarreness of his proposal, the entrant nevertheless thoughtfully points out that our expanding world population has finite quantities of space, water, atmosphere, and other resources left to provide support. By using technology to *reduce* the size of humans by two-thirds, we would extend earth's resources by a factor of 27. Smaller everything would certainly make things last longer.

Gregory Dunningham, 19 Charfield Road, Hamilton Road, Reading, Berkshire, England

THE LONGEST SOUTH-TO-NORTH WALK

The project involves a solo trek, being the furthest a person can walk from one pole to the other without crossing any large body of water. Covering some 10,000 miles, and taking $3\frac{1}{2}$ years, the objective is to walk from Punta Arenas (Tierra del Fuego) to Stanton, in the Mackenzie Territory, northern Canada, to note and record the variety of lifestyles on two continents.

Hans Walter Fricke, Max Planck Institute, D-8131 Seewiesen, Federal Republic of West Germany

NERITICA—MANNED UNDERWATER RESEARCH STATION IN THE RED SEA

This proposal involves the design and building of a relatively inexpensive manned underwater research station for use in the coral reefs of the Gulf of Aqaba, to function in the neritic region, that underwater area above the shallow continental shelf of the ocean. Its purpose is underwater research and the beginnings of human adaptation to "living" in the sea.

John D. Bush, 92 Magazine Street, Cambridge, Massachusetts 02139, U.S.A.

DID THE INCAS USE SOLAR ENERGY LONG BEFORE US?

This project seeks to explore the theory that the intricately fitted granite stones used by the Incas in making their walls were sculpted by spalling—using solar reflectors to concentrate energy in such a way as to allow the precise sculpting of the pieces to be fitted into the walls (a technique that may also have been used by the Egyptians for the casing stones on some pyramids). The theory is to be tested by actually attempting to build a section of "Incan wall."

Frank Heynick, Chopinlaan 20, Groningen, The Netherlands

LANGUAGE BEHAVIOR DURING DREAMS AND OTHER SLEEP PERIODS

Recent development of psycholinguistic models of verbal behavior has taught us much about our mental abilities and habits in the waking state. This project has been investigating the linguistic behavior of a large group of subjects who are sleep*talkers,* to determine what meaningful differences, if any, exist between our waking (conscious) and sleeping (unconscious) states. It could produce valuable new insights into the working of our brains.

John Gripenstraw, 540 13th Avenue, No. 8, Santa Cruz, California 95062, U.S.A.

INDIVIDUAL REMINDERS TO CONSERVE WATER

In drought areas, normal procedures call for public appeals to the population to save water. In this intriguing invention, the point is brought home at the individual level through the use of small, simple, inexpensive water meters for direct attachment to individual faucets and fixtures. Registering the water usage on each occasion, they create an immediate sense of conservation and a constant challenge to do better.

Narayan Ram Mayya, 8th Dwarka Kunj, Plot 509, 12th Road, Chembur, Bombay 400071, India

THE REMARKABLE MAYYA FLEXILINER DRAWING INSTRUMENT

This invention contains a built-in mechanical computing system that continuously manipulates the orientation of a template to allow the drawing of increasing- and decreasing-distance parallel lines and perspective lines, three-dimensional effect lines, and a wide variety of curves with increasing or decreasing distances. It won India's Invention Award in 1975.

Alceo Magnanini, Estrada da Vista Chinesa No. 741, Alto da Boa Vista, Rio de Janeiro, Caixa Postal 23011, ZC-08, Brazil

CAPTIVE BREEDING OF LION MARMOSETS

The rarest and most endangered of all the New World monkeys, the lion marmoset, has been given a chance to survive through captive breeding and management techniques in the Poco das Antas Biological Reserve for *Leontopithecus,* in Brazil. The three subspecies covered in this research program are slowly making a comeback from what appeared to be inevitable extinction.

William J. Spencer, 1407 Sagebrush Trail Southeast, Albuquerque, New Mexico 87123, U.S.A.

HIGH-TECHNOLOGY CONTROL OF DIABETES MELLITUS

The development of a miniature, low-power, implantable insulin delivery system for the control of diabetes mellitus, consisting of a highly accurate crystal controlled timer that actuates a piezoelectric pump and valves to deliver insulin in microliter volumes from a refillable reservoir. The system would be controlled by a self-contained programable memory or by coupling with an external programer.

Jean-Pierre L. G. L. Farcy, 12 Blvd. Henri Sappia, Le Jura, 06100 Nice, France

CAVE RESEARCH AND DEPTH RECORD ATTEMPT IN IRAN

This project is a 16-person expedition to study and collect fauna from Iranian caverns and underground rivers and lakes and to plunge, with aqualungs, into the terminal sump of Gouffre Parau at 751 meters below the entrance. This attempt may beat the previous world record of diving into an abyss because of the geological nature of the massif where the Gouffre Parau is located.

Gerald Gerard Bracken, Harbour House, Westport, County Mayo, Ireland

IS THE COMMON MARKET REALLY SOMETHING NEW?

Five thousand years ago, Europe's megalithic culture spread from Spain to Scandinavia and from Ireland to the Alps. The uniformity of the megalithic remains suggests a high degree of communication over a vast area in these ancient times. This entrant proposes to continue his work of mapping, with photography, from his own low-flying aircraft, the direct tracks between these ancient megaliths, in an effort to locate intermediate stations proving the fact of an ancient "Common Market."

Bernd Beyer, Muhlbachstrasse 2, D-7590 Achern/Baden, Federal Republic of West Germany

COULD IT BE PERPETUAL MOTION?

This entrant has developed what he calls a *gravity generator power unit*, based on a sloping piece of rail equipment, holding any number of mobile dynamo–electric direct-current generators rolling down the rails, and a gear-wheel hoisting apparatus to return the generators to the top once again. Humanity has tried to cheat nature out of energy for many years. . . .

Jacob Aljanati, Florida 2613, San Andres, F.C.B.M., Argentina

THE FUTURE PORT OF BUENOS AIRES

The vastly overloaded Port of Buenos Aires accounts for about 90 percent of Argentina's commercial trade and currently represents a severe problem for the country's economic development. This entrant, a retired merchant marine captain, has singlehandedly developed a plan for the dredging of two new channels in the port area, which would provide a new landfill city and port facilities of great value. On his own, he has come near to achieving official sanction to get the job done.

William E. Taylor, P.O. Box 116, Route No. 2, Escanaba, Michigan 49829, U.S.A.

TRACKING THE WHOOPING CRANE VIA VOICE PRINTER

Through the development and adaptation of the technology of voice printing (a reliable method for identifying human beings), this project seeks to establish a technique for identifying individual whooping cranes and greater sandhill cranes by their voices. This would be invaluable in the management and protection of these endangered species.

Daniel Sundersigh Charles Devadhar, 7 Kowhai Street, Hawera, New Zealand

NEAR ELIMINATION OF BLOOD LOSS IN SURGICAL OPERATIONS

This surgeon has developed special clamps that devascularize the uterus, which may then be rapidly cut out, with no blood loss and in about a quarter the normal time. Development of similar types of clamps for use in operating on other organs would be of great help in underdeveloped countries, where blood transfusions are difficult or impossible because of no available blood supplies or because of religious resistance to the use of transfusions.

Hideo Uchida, 4-18-5 Kamikitazawa, Setagaya-ku, Tokyo 156, Japan

A WORKING FLYING SAUCER

This project is based on the entrant's finding of a new electromotive force phenomenon, which corresponds only to an electric field and has no relationship to a magnetic field. Involving the use of an asymmetrical electrical field, a working flying saucer, 1.3 meters in diameter and 225 grams in weight, has been demonstrated. It appears that there is a direct correlation between certain radio-astronomy effects and the behavior of the saucer in flight.

Richard Stanley Bufton, Fron Haul, Abersoch, Pwllheli, Gwynedd, North Wales, United Kingdom

SEARCHING FOR A LEGENDARY GOLD SHIP

Age-old stories in Wales tell of a Spanish treasure ship, the *Santa Cruz*, which was wrecked on January 11, 1679 in Tremadoc Bay, North Wales, with a vast cargo of gold and silver. The entrant's careful research into the oral history of Tremadoc Bay has yielded clues to the area of the sinking. He proposes to use sophisticated high-technology techniques to locate, survey, and salvage the vessel.

Osvaldo L. Tonon, Libertad No. 762, Haedo, Buenos Aires, Republic of Argentina

A NEW STRUCTURAL FORM FOR ARCHITECTURE—"FOLDINGS"

The covering, or enclosure, of space is often a temporary requirement that nevertheless is approached with permanent building techniques. This entrant's development of a new mathematical-geometrical system of utilizing stress forces allows folding structures capable of spanning up to 10 meters with 2-mm thick cardboard. He has invented a wide range of structures for use in myriad applications.

Wilfred Tatham, "Vangs-Vatnet," Edies Lane, Leavenheath, Colchester, Essex CO6 4PA, England

CONSTRUCTING PERFECT VIOLINS

This development is a new concept in the construction of modern violins, in which applied electronics is used to formulate readings by a simple apparatus. The necessary notes of the natural wood are thus secured in each plate of a violin under construction, to produce the desired unison *every* time a violin is constructed, thus eliminating chance for amateur and expert alike.

Elena Beatriz Decima Zamecnik, Tokio 29, Colonia Juarez, Mexico 6 D.F., Mexico

THE VICUS CULTURE—AN UNKNOWN CROSSROAD BETWEEN NORTH AND SOUTH

The discovery, in 1963, of a previously unknown prehistoric culture, now called the Vicus, in the Piura region of northwest Peru opened the possibility that a much more ancient civilization had existed in this area than suspected. This expedition's objective is to survey and excavate these sites, to determine the possible role this society played between cultures of the Andes and cultures further north.

Cyril Leak, 4 Warwick Street, Leederville, Western Australia 6007, Australia

AMPLIFYING THE HORSEPOWER OF A LOW-POWER SOURCE

The energy crisis has forced us to examine current power sources, such as low-powered steam turbines generated by solar heat, Pelton wheels, and so on. This entrant's invention is a novel means of extracting an extra amount of horsepower from such sources, using his own design for flexible, elliptical flywheels, connected with springs, which add momentum to the prime mover's power.

Ryusuke Nagata, c/o Hando-So, 3-23-4 Minamidai, Nakano-ku, Tokyo, 164 Japan

IMPROVED METHODS FOR THE AQUACULTURE OF THE RIVER LOBSTER (PRAWN)

Despite great interest in Southeast Asia in attempting to breed prawns for food, yield has typically been very low because of many difficulties in culturing these creatures artificially. This entrant has solved the problem on a small scale and proposes to share his results and techniques with others who are interested in the process.

Harry Silsby Brown, 515 E. Micheltorena Street, Santa Barbara, California 93101, U.S.A.

FLYING EYE SURGEONS HELP THE BLIND TO SEE AGAIN

Surgical Eye Expeditions (SEE) International, founded in 1972 as an all-volunteer, humanitarian, nonprofit organization, has developed an innovative mechanism for bringing clinical eye surgeons to out-of-the-way locations to use their skills in attacking surgically correctible blindness. A portable field medical and surgical unit has been developed, and over 300 sight-restoring operations have been performed in Mexico.

Edgar Steven Wyvern Weinberg, Rijnstraat 228-3, NL-Amsterdam-1010, The Netherlands

PROTECTING THE CORAL REEFS

It has long been believed that coral reefs, among the richest biotic communities on earth, were restricted to warm seas and shallow water. The need for shallow water was attributed to the need for sunlight of the symbiotic algae associated with most corals. This project pursues the entrant's finding that the corals themselves appear to depend on light, a major breakthrough in aiding their ecological protection.

Basil Allen Barker Rossi, 7D, Bel-Air Condominium, 5022 P. Burgos Street, Makati, Rizal, Philippines

MINIATURIZED WASTE RECYCLING SYSTEMS FOR THE THIRD WORLD

The project involves the miniaturization of a number of existing waste recycling processes and their integration into a single viable plant, which would be suitable primarily for use in developing countries or underpopulated areas. The initial items under consideration include lubricating oil, silver, detinning (of tin cans), and rag (for use in the wiper trade), all at approximately 10 percent of conventional levels.

Otto Pulch, Holderlinstrasse 21, 7513 Stutensee 3, Federal Republic of West Germany

A LOW-NOISE, CLEAN-EXHAUST AIRPLANE ENGINE

Although much progress has been made in improving land engines in an environmental sense (reducing noise, polluting exhausts, and so on), very little has been done with airplane motors. This entrant has built and is testing a 750-mm airplane motor, delivering 150 hp. It should be economical and will meet US specifications on exhaust pollution up to an altitude of 3000 meters, plus being significantly quieter.

Silvia Schenkel, Dienerstrasse 28, 8004 Zurich, Switzerland

PRESERVING THE FOLK MUSIC AND DANCES OF NORTHERN GREECE

Because of the inroads of radio and television, the folk music and dances of northern Greece are disappearing from the rural areas that have nurtured them. This expedition aims to make sound recordings and film documentaries of still-existing folk music, especially children's songs, nursery rhymes, lullabies, games, and dances and to provide an ethnomusicological evaluation of their sources from this area.

Charles Glen Wilson, Zoo Director, Overton Park Zoo and Aquarium, Memphis, Tennessee 38112, U.S.A.

SEARCHING FOR THE WORLD'S RAREST PRIMATE

The Peruvian yellow-tailed woolly monkey (*Lagothrix flavicaudate*) was first discovered in 1802 in South America and rediscovered 123 years later when two different museum expeditions collected five dead specimens. One live specimen now exists in the Lima zoo. This expedition is designed to search for, locate, and study the monkey, to protect and possibly manage a captive breeding program for it.

Mohammed Mohideen Fazal Mahmood, Medical Laboratory, No. 5, First Cross Street, Chilaw, Sri Lanka

THE EARLY DIAGNOSIS OF MALARIA AND TUBERCULOSIS AT THE COMMUNITY LEVEL

This prototype program, involving about 40 villages in the general area of the entrant's laboratory, seeks to devise means whereby the combination of available laboratory facilities and social worker cooperation will enable the early detection and treatment of malaria and tuberculosis. Given local success with the method and organization, it is planned to expand the program nationally.

Gail Patricia Silverman, 3 rue Scipion, 75005 Paris, France

EXPLORING INCAN AND MAYAN TRAVELS VIA WEAVING MOTIFS

The study of migration and trade routes constitutes an important part of anthropology today, and a key aid in this study comes from comparison of artistic motifs separated by time and space. This project is concerned with comparing weaving motifs from three different weaving villages in the Chiapas area of Mexico and those of three villages in Peru and Bolivia, in an attempt to better understand Mayan and Incan migration and trade routes.

Derek Ernest Fordham, 66 Ashburnham Grove, Greenwich, London SE10 8UJ, England

BY DOG SLED ACROSS THE TOP OF THE WORLD

A husband-and-wife dog-sled expedition from northwestern Greenland to the north coast of Ellesmere Island, Northwest Territories, Canada. The purpose is to attempt to establish if the north coast was long ago used as a migration route by paleo-Eskimos moving eastward from their origin in the Coronation Gulf area of Canada, to Greenland. The expedition will be supported by air drops of food and fuel.

Pierre Lecomte, 2398 Parker, Berkeley, California 94704, U.S.A.

A PERSONAL, PORTABLE MONITORING DEVICE FOR DIVERS

Development of a miniaturized device, utilizing integrated pressure sensors, analog to digital converters, and microprocessors will provide sport and commercial divers with a personalized and constantly monitored computerized readout of their individual below-sea time and depths. This will give continual assessment of the decompression requirements needed by the diver over extended periods of time, to avoid either short-term or cumulative decompression problems.

Richard Frederick Swindell, Mount Gravatt C.A.E., P.O. Box 82, Mount Gravatt, Queensland 4122, Australia

MOBILE SOLAR INFORMATION CENTERS FOR AUSTRALIA

In spite of the high cost of petroleum-based energy and governmental interest in promoting the use of solar energy, learning and acceptance by the public have been slow. This project attempts to overcome that problem by establishing mobile information centers, using specially converted trucks and vans capable of carrying a group of demonstration models directly to public showings.

Werner Erich Popp, Fasanenweg 16, D-2303 Gettorf, Federal Republic of West Germany

CHECKING ASTROLOGICAL AND COSMOBIOLOGICAL THEORIES BY COMPUTER

Both astrology and cosmobiology have been used to predict future events, based on empirical knowledge. This project is designed to obtain massive objective data from available files (registrar's offices, police, hospital, and so on) and to analyze by computer the predictability of divorces, deaths, accidents, marriage dates, and so on.

Sally Dorothea Smith, 7 Richard Court, Alston Road, High Barnet, Herts, England

HIGH-ALTITUDE FREEFALL PARACHUTE DESCENTS—RESEARCH AND RECORDS

Objectives include establishing the world height record for civilian freefall parachuting, the woman's world height record, the United Kingdom height record, and the assessment of the maneuverability of the freefalling human body in rarefied air between the heights of 20,000 and 43,000 feet above the earth, plus developing and researching suitable equipment for these projects.

Neile Palmer, P.O. Box 20169, St. Petersburg, Florida 33734, U.S.A.

WORLD BUTTERFLY HEADQUARTERS

This project proposes the creation of a large World Lepidoptera Zoo, where live butterflies and moths can be raised and displayed for public viewing and where endangered species can be protected and bred, to prevent their extinction. Similar to an aviary, the zoo is planned to provide the necessary ecological environments for the creatures, with walkways and landscaping designed for educational and enjoyable public viewing, along with appropriate research facilities.

Ana T. Cook, Mariano H. Cornejo No. 1822, Pueblo Libre, Lima 1, Peru

SOUTH AMERICAN MAPS FOR INTERNATIONAL EXPLORERS

More than a few tragic deaths of explorers in South American expeditions have been attributed to faulty or poor cartographic information. This project seeks to produce high-quality, reliable road, trail, and travel maps, designed to assist explorers, scientists, and sport enthusiasts in planning their expeditions. Four maps are envisaged for covering the continent in a standardized information format, with hoped-for cooperation from South American governments.

William H. K. Kama, 85497 Waianae Valley Road, Waianae, Hawaii 96792, U.S.A.

ELECTRICAL ENERGY FROM DEEP-SEA CURRENTS

The objective of this project is to harness the tremendous pressures available at deep-sea current levels for the production of low-cost electrical energy. A Kaplan turbine system housed in a submersible Venturi unit is the basic means for accomplishing the metamorphosis of power. The depth of the generating system provides the needed cooling and protects it from storm damage.

Kohsei Hata, 2-3-44, Matsuzaki-machi, Osaka-fu, Osaka-shi, Japan

ELIMINATION OF HEAVY OIL BY MICROBES

Oil pollution of seas and waterways is a problem needing no introduction. This entrant has found a single strain of a bacteria (genus *Pseudomonas*) that assimilates heavy and fuel oils even in severe marine water pollution. It seems likely to be an ideal medium for combating oil spills and for cleansing oil tanks. The latter process alone is now annually responsible for about 20 million tons of oil flowing into the seas.

R. M. John Brannigan, 84 Hatton Drive, Belfast BT6 9BB, Northern Ireland

HUMAN ENDEAVOR AGAINST THE ODDS

A wife nominates a man who, at the age of two, fell and injured his spine, resulting in his confinement in a hospital until aged seven and in the prognosis that he would never walk again. A later injury, again to the spine, confined him to a wheelchair, perhaps forever. In spite of these setbacks, his singing and guitar playing led to his invitation to join a leading Irish rock group, dubbed "Ironsides" in his honor.

Jeyar Retnam Thillaimuthu, No. 2, Lorong Beluntas Satu, Medan Damansara, Kuala Lumpur

THE SHARED-ENTITY CONCEPT: A NEW LOOK AT LIGHT AND OPTICS

The entrant has made an unusual discovery in the area of light and radiation phenomena leading to the view that light does *not* disperse within optically denser media, as has been tacitly accepted. The implications are to be the subject of a book, and they provide the basis for the design and manufacture of totally achromatic lenses and systems, in addition to new areas of research in light and radiation.

Richard W. Kaiss-Chapman, MSU/ERDA Plant Research Laboratory, East Lansing, Michigan, U.S.A.

THE EFFECTS OF HORMONE LEVELS ON THE EFFICIENCY OF NITROGEN FIXATION IN PLANTS

The production of ammonia fertilizer for agricultural needs is extremely energy-expensive. A better solution may be the development of techniques that increase certain plant hormones to levels at which they can convert nitrogen in the atmosphere to a form that may be used by green plants, as this research project proposes.

Christopher John Wright, 66 Captain French Lane, Kendal, Cumbria LA9 4HP, England

A WALK ACROSS EURASIA

The entrant is an experienced solo hiker, both in England and in Europe, whose project is a journey by foot from Cape Finisterre, the westernmost point on the Iberian Peninsula, across Europe and Asia to Shanghai, on the east coast of China. A distance of about 4000 miles, the trip is estimated to last about two years, allowing for the vagaries of weather and possible difficulties of access to certain countries.

Somchit Pongpangan, ASRCT 196 Paholyothin Road, Bangkhen, Bangkok 9, Thailand

USING WASTE PRODUCTS TO REFOREST A COUNTRY

Forests in Thailand are being rapidly depleted, and current reforestation programs are not sufficient. Using coconut fiber dust, a polluting waste product of coconut decortication plants, this entrant has developed a planting medium using the dust, which can be manufactured into small blocks containing seedlings. Preliminary testing results have led to governmental interest in pilot plants and further development.

Robert L. Kuksuk, 2216 N. Lavergne Avenue, Chicago, Illinois 60639, U.S.A.

WATER QUALITY IN THE FOX RIVER VALLEY

This chemical analytical study of lakes and rivers in a concentrated residential and recreational area in the Chicago region is designed to determine, through standard laboratory and field tests, the quality of the water in the Fox River Basin and to provide useful training and experience for the teenage students of the State Microscopial Society of Illinois who are conducting the work.

Charles Kasiel Bliss, P.O. Box 222, Coogee, Sydney, Australia 2034

"BLISSYMBOLS" FOR SAVING LIVES ON THE ROAD

The invention, based on the applied sciences of linguistics, logic, and mathematics, of a universally applicable sign language capable of communicating road dangers, through illustrations, to people of any nationality, is proposed for international adoption. Additionally, the system allows for the communication of simple sentences of instruction for the traveler, thus avoiding language barriers and problems in such central locations as rail and bus stations and airports.

Antonio Betancourt del Nogal, Av. Principal Colinas de Bello Monte, Edf. el Cigarral 2-C, Caracas 105, Venezuela

THE SQUARING-CIRCLE SPIRAL GRAPH

This invention is a drawing instrument for tracing special-equation spiral curves, useful in designing highways and canals, helicoidal stairs and ramps in architecture, and in mechanical, geometric, and artistic drawing. The instrument also has the capability of "squaring the circle," thus giving a mechanical solution to a mathematical puzzle of interest for centuries and one without a solution using straightedge and compass alone.

John Call Cook, 11008 Glen Echo Court, Dallas, Texas 75238, U.S.A.

EXPLORING UNDERGROUND SPACE WITH RADAR

This project seeks to extend the entrant's previous work in forecasting for mining and tunneling projects, to predict serious hazards far enough in advance to avoid catastrophes in coal mining and tunneling. The objective is to improve the sensitivity of equipment, developed by the entrant and known as *monocycle radar,* that has been of interest internationally in countries where mining is a key part of local economies.

Wavell Frederick Cowan, 36 Parkside Street, Montreal West, Quebec
H4X 1E6, Canada

THE HANDCRAFTED PAPER PROJECT

The elegant and flourishing handmade paper industry of the eighteenth century gave way to modern machine methods, resulting in the loss of a rich human craft. This entrant has brought back the means for individuals to make their own handcrafted paper, as elegantly and as personally as imagination allows. His equipment for the process provides the opportunity for paper making to be an art form once again.

Virginia Steen-McIntyre, P.O. Box 1167, Idaho Springs, Colorado 80452, U.S.A.

HOMO SAPIENS IN THE WESTERN HEMISPHERE FOR 250,000 YEARS?

The detailed examination of the sediment samples and stabilized sediment columns (monoliths) from Hueyatlaco, Mexico, a controversial archaeological site, has uncovered well-made bifacial tools attributed to *Homo sapiens*. Recent geological evidence, including three radiometric dates, indicates that the bones of animals associated with the artifacts, together with the sediments that contain them, are approximately 250,000 years old. Confirmation of this would be startling news.

David Oliver Lavallee, 43 Blvd. Frederic Mistral, Toulon 83100, France

THE ATLANTIC DRIFT PROJECT

The objective is to collect data on marine biota from the air–sea interface to a depth of 100 feet, using a specially designed surface platform. It will permit subsurface viewing and extended (90 to 120 days) on-station living for scientific observers, who will drift with the vehicle, after its launching, with the flow of the Gulf Stream and the North Atlantic currents. Ecological data would be used as a base for comparison in future pollution research.

William Edward Wallace, Department of Chemistry, University of Pittsburgh, Pittsburgh, Pennsylvania 15260, U.S.A.

SOLAR ENERGY TO POWER AN ELECTRIC VEHICLE

This development of means to capture and store solar energy and to use it for vehicular propulsion involves using certain rare earth intermetallics demonstrated by the entrant to be exceedingly effective in ammonia synthesis and other reactions, as basis for the manufacture of hydrogen/air fuel cells of novel design as electrode catalysts. Propulsion will derive from recently developed compact electric motors.

Richard Eric Rex Haylock, The New House, Fox Lane, Kirby Muxloe, Leicester, LE9 9AG, England

ATTEMPTING TO SAVE THE BARBARY LEOPARD

This is an expedition to find, film, and estimate the numbers of Barbary leopards in the Mid-Atlas Mountains of Morocco. If possible, it will capture and return a small number of breeding animals for captive breeding management, in order to prevent the probable extinction of this isolated subspecies, only some 50–100 of which are believed to exist.

Mahlon Cooper Smith, 225F Engineering Building, Michigan State University, East Lansing, Michigan 48824, U.S.A.

SOLO DIRIGIBLE FLIGHT OVER THE ATLANTIC

The project is the designing, building, and flying of a small, one-person dirigible across the Atlantic, in the spirit of the solo flight of Lindbergh and the solo sea voyages of Chichester, to illustrate the feasibility of this form of transportation. It is superior to the balloons used recently in attempted crossings because of the power and steering capabilities of the dirigible.

NAME INDEX

Aljanati, J., Argentina, 345
Alvarez, R., Mexico, 339
Anderson, D. C., Scotland, 338
Anderson, M., England, 338
Araoz, A. B., Argentina, 340
Arnold, R. M., U.S.A., 246
Asmus, J. F., U.S.A., 210

Balek, J., Czechoslovakia, 160
Baur, H. F., Switzerland, 80
Beyer, B., West Germany, 344
Bliss, C. K., Australia, 356
Bocks, J. C., England, 250
Bowin, K. E., Sweden, 340
Bracken, G. G., Ireland, 344
Brannigan, R. M. J., Ireland, 354
Brooke, B. J., England, 294
Brown, H. S., U.S.A., 348
Buck, J. S. M., Argentina, 72
Bufton, R. S., North Wales, 346
Bulley, O. M., England, 322
Bush, J. D., U.S.A., 342

Chandran, T. C., India, 339
Churchill, T., England, 198
Clarkson, G. W. H., England, 316
Cohen, N. W., U.S.A., 170
Cole, M. E., England, 24
Cook, A. T., Peru, 353
Cook, J. C., U.S.A., 356
Cowan, W. F., Canada, 357

Debecker, L. J-F., Belgium, 12
Delamare, G. M. A., France, 130
Deltgen, F., Germany, 118
Devadhar, D. S. C., New Zealand, 345
Dujardin, B. A. F., France, 150
Dunningham, G., England, 341

Epstein, W., Canada, 88

Farcy, J-P. L. G. L., France, 344
Faulin, A., Italy, 240
Fedorcsak, I., Hungary, 320
Fentress, J. C., U.S.A., 138
Fernando, A., Australia, 188
Fordham, D. E., England, 351
Frank, J. H., England, 66
Franklin, D., U.S.A., 18
Fricke, H. W., West Germany, 341
Fujii, H., Japan, 178

Gaylord, T. K., U.S.A., 284
Gilead, E., Israel, 340
Gill, L. H., U.S.A., 62
Giovando, G., Italy, 8
Goodman, R. M., U.S.A., 228
Greenamyer, D. G., U.S.A., 34
Gripenstraw, J., U.S.A., 342
Guichard, C. P., U.S.A., 266

Hata, K., Japan, 353
Haylock, R. E. R., England, 358
Hendrickson, P. H., U.S.A., 114
Heynick, F., Netherlands, 342
Hoeck, H. N., Colombia, 280

Kaiss-Chapman, R. W., U.S.A., 354
Kama, W. H. K., U.S.A., 353
Knisley, M. R. M., U.S.A., 341
Kristen, H., Austria, 134
Kuksuk, R. L., U.S.A., 355

Lasley, B. L., U.S.A., 58
Lavallee, D. O., France, 357
Leak, C., Australia, 347
Lecomte, P., U.S.A., 351
Lord, R. D., U.S.A., 326

Magnanini, A., Brazil, 343
Mahapatra, U. P., India, 308

Mahmood, M. M. F., Sri Lanka, 350
March, A. A. C., England, 254
Marten, K. L., U.S.A., 260
Mayya, N. R., India, 343
Milne, D. L., South Africa, 224
Missongo, J-R., Congo, 164
Mohanlal, V., India, 194
Monsod, G. G., Jr., Philippines, 202

Nagata, R., Japan, 348
Noailles, C. P. N. V., Argentina, 288
del Nogal, A. B., Venezuela, 356

Olivera, V. A., Argentina, 104
Olivier, R. C. D., England, 270
Overland, M. C., U.S.A., 46

Palmer, N., U.S.A., 352
Parsons, R. R., England, 2
Patterson, F. G. P., U.S.A., 182
Pecero, F. A., Spain, 144
Pengson, L. T., Philippines, 110
Pernette, J-F., France, 52
Pomeroy, R. L., U.S.A., 124
Pongpangan, S., Thailand, 355
Popp, W. E., West Germany, 352
Pulch, O., West Germany, 349

Restanio, J. R., Argentina, 312
Revilla, M., Republic of China, 337
Romero, F. A., Spain, 220
Rossi, B. A. B., Philippines, 349

Saad, S. G., Egypt, 40
Schenkel, S., Switzerland, 349

Senanayake, L., Sri Lanka, 334
Shepard, D. A., U.S.A., 302
Sheppard, N., England, 98
Silverman, G. P., France, 350
Smith, M. C., U.S.A., 358
Smith, S. D., England, 352
Snellman, C. H., Finland, 28
Spencer, W. J., U.S.A., 343
Steen-McIntyre, V., U.S.A., 357
Swindell, R. F., Australia, 351

Tatham, W., England, 347
Taylor, W. E., U.S.A., 345
Teeuwen, G. H., Netherlands, 216
Thillaimuthu, J. R., Malaya, 354
Thompson, L. G., Australia, 84
Tonon, O. L., Argentina, 346

Uchida, H., Japan, 346

Wallace, W. E., U.S.A., 358
Walter, H., West Germany, 337
Watson, R. M., England, 94
Weinberg, E. S. W., Netherlands, 348
Westby, N., Norway, 338
Wilson, C. G., U.S.A., 350
Woodford, G. R., England, 339
Wright, C. J., England, 355

Yu, M. L., China, 234

Zamecnik, E. B. D., Mexico, 347
Zimmerman, M. R., U.S.A., 154
Zink, D. D., U.S.A., 330
Zucker, M. H., U.S.A., 276